## Proceedings of the Lebedev Physics Institute Academy of Sciences of Russia

| | |
|---|---|
| Volume 165 | **Solitons and Instantons, Operator Quantization,** Edited by V.L. Ginzburg |
| Volume 166 | **The Nonlinear Optics of Semiconductor Lasers,** Edited by N.G. Basov |
| Volume 167 | **Group Theory, Gravitation and Elementary Particle Physics,** Edited by A.A. Komar |
| Volume 168 | **Issues in Intense-Field Quantum Electrodynamics,** Edited by V.L. Ginzburg |
| Volume 169 | **The Physical Effects in the Gravitational Field Black Holes,** Edited by M.A. Markov |
| Volume 170 | **The Theory of Target Compression by Longwave Laser Emission,** Edited by G.V. Sklizkov |
| Volume 171 | **Research on Laser Theory,** Edited by A.N. Orayevskiy |
| Volume 172 | **Phase Conjugation of Laser Emission,** Edited by N.G. Basov |
| Volume 173 | **Quantum Mechanics and Statistical Methods,** Edited by M.M. Sushchinskiy |
| Volume 174 | **Nonequilibrium Superconductivity,** Edited by V.L.Ginzburg |
| Volume 174 Supplemental Volume 1 | **Thermodynamics and Electrodynamics of Superconductors,** Edited by V.L. Ginzburg |
| Volume 174 Supplemental Volume 2 | **Electron Liquid Theory of Normal Metals,** Edited by V.P. Silin |
| Volume 175 | **Luminescence Centers of Rare Earth Ions in Crystal Phosphors,** Edited by M.D. Galanin |

| | |
|---|---|
| Volume 175 Supplemental Volume | **Luminescence and Anisotropy of Zinc Sulfide Crystals,** Edited by M.D. Galanin |
| Volume 176 | **Classical and Quantum Effects in Electrodynamics,** Edited by A.A. Komar |
| Volume 176 Supplemental Volume | **Quantization, Gravitation and Group Methods in Physics,** Edited by A.A. Komar |
| Volume 177 | **Stoichiometry in Crystal Compounds and Its Influence on Their Physical Properties,** Edited by Yu. M. Popov |
| Volume 178 | **The Delfin Laser Thermonuclear Installation: Operational Complex and Future Directions,** Edited by G.V. Sklizkov |
| Volume 179 | **X-Ray Plasma Spectroscopy and the Properties of Multiply-Charged Ions,** Edited by I.I. Sobel'man |
| Volume 179 Supplemental Volume 1 | **Interaction of Ultrashort Pulses with Matter,** Edited by M.D. Galanin |
| Volume 179 Supplemental Volume 2 | **Electron-Excited Molecules in Nonequilibrium Plasma,** Edited by N.N. Sobolev |
| Volume 180 | **Inelastic Light Scattering in Crystals,** Edited by M.M. Sushchinskiy |
| Volume 181 | **Pulsed Metal and Metal Halogenide Vapor Lasers,** Edited by G.G. Petrash |
| Volume 182 | **Luminescence of Wideband Semiconductors,** Edited by M.D. Galanin |
| Volume 183 | **Invariants and the Evolution of Nonstationary Quantum Systems,** Edited by M.A. Markov |
| Volume 184 | **Electron Processes in MIS-Structure Memories,** Edited by A.F. Plotnikov |

Volume 185 **Injection Lasers in Optical Communication and Information Processing Systems,** Edited by Yu. M. Popov

Volume 186 **Photoproduction of Pions on Nucleons and Nuclei,** Edited by A.A. Komar

Volume 187 **Quantum Field Theory, Quantum Mechanics and Quantum Optics: Pt. 1. Symmetries and Algebraic Structures in Physics,** Edited by V.V. Dodonov and V.I. Man'ko

Volume 188 **Integral Systems, Solid State Physics and Theory of Phase Transitions. Pt. 2. Symmetries and Algebraic Structures in Physics,** Edited by V.V. Dodonov and V.I. Man'ko

Volume 189 **Clusters of Galaxies and Extragalactic Radio Sources,** Edited by A.D. Kuz'min

Volume 190 **Metal-Optics and Superconductivity,** Edited by A.I. Golovashkin

Volume 190 Supplemental Volume **Superconductors with A15 Lattice and Bridge Contacts Based Upon Them,** Edited by M. M. Sushchinskiy

Volume 191 **Polarization of the Vacuum and a Quantum Relativistic Gas in an External Field,** A.Ye. Shabad

Volume 192 **Atomic and Ionic Spectra and Elementary Processes in Plasma,** Edited by I.I. Sobelman

Volume 193 **Research on Chemical Lasers,** Edited by A.N. Orayevskiy

Volume 194 **Biophysical Approach to Complex Biological Phenomena,** Edited by E. Volkov

Volume 195 **Physical Characteristics and Critical Temperature of High Temperature Superconductors,** Edited by M.M. Sushchinskiy

Volume 196 **Pulsars,** Edited by A.D. Kuzmin

Volume 197 **Reliability Problems of Semiconductor Lasers,** P.G. Eliseev

Volume 198 **Theory of Nonstationary Quantum Oscillators,** Edited by M.A. Markov

| | |
|---|---|
| Volume 199 | **Spatiotemporal Characteristics of Laser Emission,** M.V. Pyatakhin and A.F. Suchkov |
| Volume 201 | **Generation of Nonlinear Waves and Quasistationary Currents in Plasmas,** Edited by L.M. Kovrizhnykh |
| Volume 202 | **Gas Dynamics,** Edited by Yu.I. Koptev |
| Volume 203 | **Fiber Optics: Research and Development,** Edited by Ye.M. Dianov |
| Volume 204 | **Sakharov Memorial Lectures in Physics,** Edited by L.V. Keldysh and V.Ya. Fainberg |
| Volume 205 | **Squeezed and Correlated States of Quantum Systems** Edited by M.A. Markov |
| Volume 206 | **Mathematical Physics, Applied Mathematics and Informatics,** Edited by Yu.I. Koptev |
| Volume 207 | **Generalized Mayer Series in Classical Statistical Mechanics** Igor J. Ivanchik |
| Volume 208 | **Physical Properties of High Temperature Superconductors,** Edited by Yu.I. Koptev |
| Volume 210 | **Theory and Technology of High Temperature Superconductors,** Edited by Yu.I. Koptev |
| Volume 216 | **Semiconductors and Insulators: Optical and Spectroscopic Research,** Edited by Yu.I. Koptev |
| Volume 217 | **Nonlinear and Quantum Optical Phenomena in Nonequilibrium Media,** Edited by V.A. Shcheglov |
| Volume 218 | **Radiative and Collisional Characteristics of Ions in Hot Plasmas,** Edited by I.I. Sobel'man |
| Volume 219 | **The Muon Method in Science** V.P. Smilga and Yu.M. Belousov |

Proceedings of the Lebedev Physics Institute
Academy of Sciences of Russia
Volume 217

# NONLINEAR AND QUANTUM OPTICAL PHENOMENA IN NONEQUILIBRIUM MEDIA

Edited by
V.A. Shcheglov

Nova Science Publishers, Inc.

Deputy Series Editor: M.A. Man'ko

Art Director: Christopher Concannon
Graphics: Elenor Kallberg and Maria Ester Hawrys
Book Production: Michael Lyons, June Martino,
Tammy Sauter, and Michelle Lalo

ISBN 1-56072-138-3

*Library of Congress*
*Cataloging-in-Publication Data*
*available upon request*


*6080 Jericho Turnpike, Suite 207*
*Commack, New York 11725*
*Tele. 516-499-3103 Fax 516-499-3146*
*E Mail Novasci1@aol.com*



*Printed in the United States of America*

# CONTENTS

## TRANSFORMATION OF SPATIAL STRUCTURE OF MONOCHROMATIC RADIATION IN NONLINEAR MEDIA

***D.Yu. Kuznetsov***

## THEORY OF DEGENERATE INTERACTION OF MULTIFREQUENCY IR RADIATION IN RESONANT MEDIA

*I.M. Bel'dyugin, A.A. Stepanov and V.A. Shcheglov*

## EFFECT OF NONMONOCHROMATICITY OF OPTICAL BEAMS ON THE EFFICIENCY AND QUALITY OF FOUR-WAVE INTERACTION

*A.I. Erokhin, A.A. Stepanov and V.A. Shcheglov*

# TRANSFORMATION OF SPATIAL STRUCTURE OF MONOCHROMATIC RADIATION IN NONLINEAR MEDIA

*D.Yu. Kuznetsov*

The purpose of this paper is analytical and numerical investigation of field spatial structure distortion effects associated with an amplifying medium nonlinearity.

Actual need of theoretical analysis of a spatially inhomogeneous radiation in an amplifying medium is conditioned by a number of factors of practical importance. The need of spatial beam structure investigation in high-power laser systems, signal distortion problem in brightness amplifiers, laser data-handling system development and active formation of a predetermined spatial field structure should be noted here. In such systems, nonlinearity is a point of principle, since inversed population of active medium levels is utilized effectively only if amplification proceeds in a saturation mode.

Intensive studies of radiation interaction with nonlinear media have been performed since the early sixties. These investigations are described in [1-4]. Nonlinear interaction between radiation and matter depends on spatial field structure. This dependence was studied in a large number of papers. In particular, in [5] interference between colliding waves is shown to decrease the efficiency of utilization of inversed population of a medium. In [6] a transformation of a beam structure with the Gaussian intensity profile in lenslike media is investigated, in [7] a method of calculation of active resonator properties is developed in approximation of Gaussian intensity distribution in a beam cross section.

More precise results are provided by a so-called moment method [8-10], its results agree with the results of direct numerical calculations [11]. The moment method allows an obvious physical interpretation based on analogy between a paraxial wave beam and a dynamic system [12]. Waveguide field configurations in a self-focusing medium [13-15] correspond to steady states of this dynamic system [12]. Its symmetry results in conservation of certain integral quantities, i.e. Noether integrals (see, for example, [16,17]). With the help of these integrals self-focusing collapse was investigated in [18] for the case of a medium with a third-order nonlinearity.

If a more detailed examination of the radiation spatial inhomogeneity effects is needed, numerical solution of nonlinear diffraction equation is carried out (see, for example, [19-22]). But numerical methods are efficient only for the fields with a relatively simple structure. For beams with a complex structure they are cumbersome, and application of statistical methods is advisable. In [22] comparison of numerical experiment results and statistical calculations [23] is carried out. Applicability of statistical methods of field description in a medium with a weak nonlinearity is substantiated. Analytical approach [24] should also be noted but it has been developed only for several particular cases. In particular, transformation of correlation properties of a complex field in nonlinear (and also in randomly inhomogeneous) media was studied in [25,26] using statistical averaging. A similar method was used in [27-36] dealing with phase conjugation and holography and also in investigation of propagation of intense beams in the atmosphere [27, 37,38].

Statistical approach was used in [39] for a purely active medium (amplification in a saturation regime). Nonlinear distortions of a beam with a partial spatial modulation and low diffraction length are shown to be amenable to analytic estimations and consist primarily of a modulation depth decrease.

Generalization of this investigation to the case of a medium with a nonlinear refraction is of considerable interest. It is conditioned by the problem of development of high-power laser installations (perturbation growth due to self-focusing instability is one of the principal factors limiting the power density; see, for example, [40, 41]), systems with laser processing of both information [52-55] and material objects [53,56], and also systems with active formation of a field with a predetermined spatial structure [57,60]. This generalization is the subject of this publication. It is based upon the results of [61-69] and has the following structure.

In Ch. 1 equations are derived that describe interaction of colliding optical beams in paraxial approximation. The case of a single paraxial beam is considered at greater length, and general relations are formulated that do not require special assumptions concerning the features of the transverse field structure. The influence of amplification on the beam self-focusing instability is discussed, applicability of the decay integral (see, for instance, [40,41]) for the estimation of an extent of a plane wave perturbation growth due to self-focusing instability is considered.

In Ch. 2 a wide partially modulated beam is considered in a medium wiht a weak nonlinearity of general type. In the frameworks of the speckle-mode theory, approximate solution of the diffraction equation is constructed and its uncertainty is estimated. Conclusion is drawn concerning the regularities of the field distortions in the case of a weak nonlinearity. The conclusions of Ch. 1 about the contrast transformation resulting from interaction between a modulated field component and a plane wave are generalized to the case of a finite (not infinitesimal) modulation depth.

In Ch. 3 coupled wave equations are examined for the case of a saturation type nonlinearity and for a third-order nonlinearity. Nimerical estimates of image contrast transformation in physical media are made. A technique for measuring nonlinear parameters of a media, in particular, nonlinear refraction constants, is proposed, which is based on a sharp dependence of a contrast on a phase mismatch. A numerical experiment is described that confirms the conclusion about regularities of nonlinear distortion of a complex field.

In Ch. 4 interaction between colliding light beams in an extended amplifier is explored. The results obtained are applied to the calculations of a double-pass amplifier. The features of degenerate four-wave interaction appearing in nonlinear amplification of colliding partially modulated beams are analyzed, and their practical importance is studied.

Principal conclusions are summarized in the end

# Chapter 1

# Propagation of radiation in a medium

In this chapter applicability of nonlinear Helmholtz equation to actual physical media is analyzed, concept of complex increment is introduced and possible magnitudes of this quantity are considered. For paraxial propagation of radiation, reduced equations are presented and their general properties are examined. As a preliminary presentation of the foregoing material, weak perturbations on a plane wave background in an amplifying medium with nonlinear refraction are considered.

## 1.1. Actual media

In this section the nonlinear Helmholtz wave equation is written for monochromatic polarized radiation in an active medium. Concept of local increment is introduced. Locality of intensity dependence of this quantity is analyzed. Possible magnitudes of the local increment in actual physical media are considered.

Let us assume that beams with a definite polarization are propagating in a nonlinear (active) medium, and the medium properties are determined by two real quantities, namely, refractive index $n(I) = n_0 + \tilde{n}(I)$ and local increment $\bar{g}(I)$ where $I = E_c E_c^*$ is the modulus squared of the field complex amplitude $E_c \exp(-i\omega t)$ in the same sence as, for example, in [36]. It is helpful to combine the two real quantities into one defining a complex increment $g(I) = \bar{g}(I) + ik\tilde{n}/n_0$ where $k = 2\pi/\lambda = \sqrt{n_0}\,\omega/c$ is a wavenumber. Using this notation we write the Helmholtz wave equation (see, for example, [4,70,71]) in the form of

$$(\Delta - k^2)\, E_c = 2ikg\,(E\ E_c^*)\, E_c. \tag{1.1}$$

It is assumed that $\varepsilon\mu\omega^2/c^2 = k^2 + ikg$, where $\omega$ is the field frequency; $\varepsilon$, $\mu$ are complex dielectric and magnetic permittivities at this frequency; $c$ is the velocity of light. We may consider the actual physical fields $E_f$ and $H_f$ to be connected with the field $E_c$ through the formulae

$$\begin{aligned} &\mathbf{E}_f = \mathrm{Re}\,(\mathbf{E}_c \exp(-i\omega t), \qquad \mathbf{H}_f = \mathrm{Re}\,(\mathbf{H}_c \exp(-i\omega t)),\\ &\mathbf{E}_c = \mathbf{e}E_c, \qquad \mathbf{H}_c = \mathbf{B}/\mu = (i\omega\mu)^{-1} c\,\mathrm{rot}\,\mathbf{E}_c, \end{aligned} \tag{1.2}$$

where $\mathbf{e}$ is the polarization vector of the electric field. We note that this expression is valid only if definite assumptions concerning the field polarization are made (for instance, if polarization is a linear one); otherwise, when magnetic field is eliminated from the Maxwell equations (see, for example, [4]), a term appears containing rot rot $E_f$, and Eq. (1.1) cannot be used.

The following treatment suggests that the field dependence of the complex increment is local in space and does not change on a scale of radiation transit time through the system, while the responce time must be smaller than the optical pulsewidth. It is important to specify what media and what fields can be a subject of discussion. In this connection we consider possible magnitudes of the local increment in actual amplifying media.

In interaction of a high-power radiation with a medium, a temperature raises, internal stresses and even damages (in solid-state media) may appear and chemical composition may vary. All these mechanisms lead to the nonlocality of the medium response both in time and space. However, for many media a range of parameter values can be found, in which the basic medium properties can be described using a complex increment that depends locally on the passing light intensity. For instance, in [72-75] nonlinear properties of such media as carbon disulphide, nitrobenzene, neodimium glass and barium chloride are successfully described by introduction of only one parameter, i.e. nonlinear refraction constant determining a direct proportionality between imaginary part of the increment and the light intensity, moreover, relative measurement uncertainty for this parameter in many cases does not exceed several percent.

In [3] conditions are discussed, when the mechanisms resulting in delocalization of the medium response under the action of light in solids and liquids are inessential. For instance, in liquids, electrostriction (leading to spatial and temporal nonlocality in general case) can be considered a local nonlinearity mechanism at $\ell_\perp v_s < \tau$, where $\ell_\perp$ is a scale length of field modulation, $v_s$ is the sound velocity, $\tau$ is a pulsewidth.

Question of response locality in gaseous media needs a special consideration. At high temperatures ($T \approx 2000$ K in a Cu vapor amplifier) atoms can change their position during the characteristic lifetime of the upper laser level. For instance, they can displace from a node of an interference pattern to its antinode. At T = 2000 K copper atoms with the mass $m \approx 10^{-22}$ g travel with velocities of the order of $v = \sqrt{T/m}$ = 50000 cm/s. For radiation with a wavelength $\lambda = 0.51 \cdot 10^{-4}$ cm (green line) Doppler broadening is $\Delta\nu = \nu v/c \simeq 10^{9}\ s^{-1}$. Homogeneous broadening is somewhat lower (see, for example, [76]), therefore, the excited state lifetime can be assumed to be $\tau_0 > 1/\Delta\nu \approx 10^{-9}$ s. Since a typical laser pulsewidth is $10^{-8}$ s, the radiation still can be considered monochromatic.

The estimate shows that the medium response can be considered local for unidirectional propagation of radiation in a Cu vapor-based active medium. In the presence of counter-propagating waves, when the characteristic scale length of the interference pattern is of the order of $\lambda/2$, the interference pattern induced in the medium by the field is partially washed out.

We note that nonmonochromaticity of light beams may also result in a loss of locality if the longitudinal relaxation time exceeds the radiation coherence time $\tau_{coh}$. In this case coherence degree of the interacting beams is an essential factor (they are supposed to come from the same source). If the path difference between these beams is $\Delta L >> c\tau_{coh}$, the medium responds only to intensity averaged over relative beam phases even in the presence of a spatially local response.

Now we proceed to the question of values taken by the local increment and relation between refractive and active nonlinearities. We note that it is reasonable to speak about variations of the absolute value of the imaginary part of the increment in dependence on the radiation intensity rather than about the absolute value itself.

The reference point is determined from the actual values of the constant part of the refractive index $n(I)$. In cases when the refractive index tends to a constant value $n(\infty)$ with the intensity growing, this very value is taken as the constant part of the refractive index. This situation is realized, for example, in dyes, when both refractive and active nonlinearities are related to the variation of populations of the working levels in the medium. If the refractive index saturation is observed only at very high intensities, then the linear refractive index is considered to be the constant part of the refractive index.

In condensed media (liquids and solids) refractive nonlinearity is usually stronger than active one. So, active nonlinearity is often neglected, when transformation of the field structure is surveyed (see, for example, [8-15,72]). For instance, in [41] operation mode of a neodimium glass amplifier is described ($\lambda = 1.06 \cdot 10^{-4}$ cm, $I \approx 10^{9}$ W/cm$^{2}$), in which the amplification coefficient was of the order of four at the amplification length $L = 54$ cm, it corresponded to $\mathrm{Re}\, g \approx 0.02$ cm$^{-1}$. Nonlinear phase shift in such an amplifier is of the order of $\pi$, and it corresponds to $\mathrm{Im}\, g \approx 0.05$ cm$^{-1}$. The contribution of the active nonlinearity increases with the active centre concentration in the medium growing [73], the situation $\mathrm{Re}\, g > \mathrm{Im}\, g$ may be realized in future. Applicability of monochromatic field equations to such media is restricted by the fact that the medium relaxation time may be longer than the optical pulsewidth (the "monochromatic" equations give only qualitative description here).

Eq.(1.2) is applicable to dyes in most cases. Relaxation time is often shorter than the pulsewidth. For instance, in [77,78] direct proportionality between amplification (absorption) coefficient and refractive index (or, in our notations, between real and imaginary parts of the increment $g$) is shown to hold for R6G and DODCI dyes in a wide range of intensity and concentration variations. In [79] image amplification at the wavelength of $\lambda = 0.57 \cdot 10^{-4}$ cm in an active medium based on R6G dye is described. The gain coefficient value of 23 dB/mm given in this work corresponds to $\mathrm{Re}\, g \approx 10$ cm$^{-1}$. From proportionality of real and imaginary parts of the increment infered in [77,78] one can conclude that $\mathrm{Im}\, g \approx 8$ cm$^{-1}$ in this case.

In gaseous media nonlinearity usually arises only from variations of the level population in a medium. At high amplification values ($g_0 L \approx 10$ and more for a copper vapor amplifier) and for amplified field frequencies in the wing of an amplification line, nonlinear phase shift can be of the order of unity. It will be shown later (Sec. 1.4) that it leads to essential distortions of an image carried by a partially modulated field.

So, nonlinearities both refractive and active can participate in image structure transformations in amplifiers. Their relative importance is determined by actual experimental conditions. It relates to a wide class of active media, so in Secs. 1.3–2.4 the material is presented for a general type of nonlinearity, if possible. Secs. 1.3 and 3.2 are specially devoted to consideration of refractive nonlinearity.

## 1.2 Colliding light beams; statement of the problem

In this section paraxial diffraction equations describing interaction of colliding paraxial beams with a nonlinear medium are derived on the basis of the nonlinear Helmholtz equation (1.1). Several particular cases of the intensity dependence of the complex increment are analyzed.

There is a wide-spread situation in optical systems when radiation is propagating preferentially along a certain direction, i.e. an optical axis. Bearing this in mind, we shall examine approximate solutions of Eq. (1.1) of the form of

$$E_c = E_1 \exp(ikz) + E_2 \exp(-ikz), \tag{1.3}$$

where $E_1$ and $E_2$ are functions with weak spatial dependence. We assume that they vary noticeably only on a scale of the order of $\ell_\perp$, such that $k\ell_\perp \gg 1$. The functions $E_1$ and $E_2$ correspond to beams travelling in z direction and in the opposite direction, respectively. Substituting (1.3) into (1.1) we obtain

$$\begin{aligned}\exp(ikz)\,(2ik\,\partial/\partial z + \Delta)\,E_1 + \exp(-ikz)\,(-2ik\partial/\partial z + \Delta)\,E_2 = \\ = 2ikg\,(|\,E_1 \exp(2ikz) + E_2\,|^2)\,(E_1 \exp(2ikz) + E_2 \exp(-ikz)).\end{aligned} \tag{1.4}$$

Since the scale of longitudinal inhomogeneities of the fields $E_1$ and $E_2$ is determined by the diffraction length $L_d = k\ell_\perp^2 \gg \ell_\perp$, the second derivatives with respect to $z$ are small as compared to the second derivatives with respect to the transverse coordinates. So, we neglect these derivatives. Then in formula (1.4) Laplacian $\Delta$ reduces to $\Delta_\perp = \Delta - \partial^2/\partial z^2 = \partial^2/\partial x^2 + \partial^2/\partial y^2$. Let us multiply (1.4) by $\exp(-ikz)/(2ik)$ and take the average over $z$ from $z - \pi/k$ to $z + \pi/k$. In view of weak dependence of $E_1$ and $E_2$ on $z$, their derivatives can be pulled outside the averaging sign in the left side (this is not possible in the right side because $g$ depends on the rapidly oscillating argument). These transformations result in the paraxial equation

$$\left(\frac{\partial}{\partial z} + \frac{\Delta_\perp}{2ik}\right) E_1 = \frac{1}{\pi} \int\limits_{-\pi/2}^{\pi/2} g\,(|\,E_1 + E_2 e^{-2ikz}\,|^2)(E_1 + E_2 e^{2ikz})\,d\,(kz). \tag{1.5}$$

We note that for linear medium ($g = 0$) this equation was derived in [80]. In a similar way (multiplying (1.3) by $-\exp(ikz)/(2ik)$ and taking the average) we obtain the equation for the counter-propagating beam

$$\left(-\frac{\partial}{\partial z}+\frac{\Delta_\perp}{2ik}\right)E_2=\frac{1}{\pi}\int\limits_{-\pi/2}^{\pi/2} g(|E_2+E_1e^{-2ikz}|^2)(E_2+E_1e^{2ikz})\,d(kz). \tag{1.6}$$

We note that the same approximation was used, in particular, in [5,81, 82] for the concrete $g(I)$ functions.

In view of slow dependence of the functions $E_1$ and $E_2$ on $z$, these functions may be considered constant when integrating in the right sides of (1.5) and (1.6). Let us define a complex function $F(I_1, I_2)$ of two real arguments $I_1$ and $I_2$, according to the formula

$$F(I_1,\ I_2)=\frac{1}{2\pi}\int\limits_{-\pi}^{\pi} g\left(\left|\sqrt{I_1}+\sqrt{I_2}\,e^{i\theta}\right|^2\right)\left(1+\sqrt{I_2/I_1}\,e^{i\theta}\right)d\theta. \tag{1.7}$$

Let us rewrite the system (1.5),(1.6) in the form of

$$\begin{aligned}(\partial/\partial z+\Delta_\perp/(2ik))\,E_1&=F\,(E_1E_1^*,\ E_2E_2^*)\,E_1,\\ (-\partial/\partial z+\Delta_\perp/(2ik))\,E_2&=F\,(E_2E_2^*,\ E_1E_1^*)\,E_2.\end{aligned} \tag{1.8}$$

To formulate a physical problem, one should impose boundary conditions for the field $E_1$ at the left border of the nonlinear medium (for definiteness, in the plane $z = 0$) and for the field $E_2$ at the right border of the medium (for definiteness, in the plane $z = L$). The function $F$ in the right sides has the sence of an increment. In a purely amplifying medium, Im $g = 0$, and the increment $F$ is real. In a purely refractive medium, Re $g = 0$, and the increment $F$ is imaginary.

Let us analyze the question of the system of units of measurements for physical quantities. Only paraxial beams are considered here, so it is convenient to measure the electric field strength in such units that the field modulus squared is equal to intensity. In Secs. 3.3, 4.3–4.5 intensity is measured in the units determining amplification saturation in the medium. This deviation from the standard intensity definition as a mean value of a modulus of the Poynting vector

$$\langle | \mathbf{S} | \rangle = \langle \, [\mathbf{E}_{\mathcal{F}}, \ \mathbf{H}_{\mathcal{F}}] \, | \, (c/(4\pi) \rangle = \sqrt{\varepsilon/\mu} \, EE^* c/(16\pi)$$

is justified by easiness of the inverse transition and convenience of wrighting the subsequent formulae.

We turn now to particular cases of dependence of $g(I)$.

Let $g(I) = i\gamma I$, $\gamma$ = const (third order nonlinearity). Then

$$F(I_1, \ I_2) = i\gamma \, (I_1 + 2I_2). \tag{1.9}$$

This is the case usually considered in the works dealing with phase conjugation (see, for example, [27-29,31].

Let $g(I) = g_0/(1 + I/I_s)$; $g_0$ = const is the unsaturated increment, $I_s$ = const is the saturation intensity. In this case

$$F(I_1, \ I_2) = \frac{g_0 I_s}{2 I_1} \left( 1 - \frac{1 - (I_1 - I_2)/I_s}{(1 + 2\,(I_1 + I_2)/I_s + (I_1 - I_2)^2/I_s^2)^{1/2}} \right). \tag{1.10}$$

This is a saturation type nonlinearity. The approximation was used in [5,30,82]. This dependence is presented in Fig. 4.7 of Ch. 4 to compare it with effective increments of modulated beams. The range $I_2 > I_1 + I_s$ may be noted, in which the increment $F(I_1, I_2)$ of the field $E_1$ grows with its intensity $I_1$. Amplification of each beam is saturated by the beam itself not so efficiently as by the counter-propagating beam. Dependence of the increment on the counter-propagating beam intensity for both beams should affect their modulation. Transformation of the modulation of the colliding beams is discussed in Ch. 4 in a greater detail.

We note that in [83, 84] dependence of the form of

$$F(I_1, I_2) = g_0 (1 + (I_1 + I_2)/I_s)^{-1}. \tag{1.11}$$

is used. Such a dependence does not hold for local responce of a medium, nevertheless, it has a physical sence. Dependence of the form of (1.11) corresponds to noninterfering colliding beams. This question is discussed in Sec. 4.5 at a greater length.

The purpose of the subsequent treatment is the study of nonlinear beam transformation in a medium on the basis of the analysis of properties of solutions of system (1.8). In particular, the local increment is $\mathcal{F}(E_1 E_1^*, 0) = g(E_1 E_1^*)$ for unidirectional mode of radiation propagation (for instance, at $E_2 = 0$), and so it is possible to change from system (1.8) to the equation

$$(\partial/\partial z + \Delta_\perp/(2ik))\, E = g\,(EE^*)\, E. \tag{1.12}$$

For purely refractive media (Re $g = 0$), Eq. (1.12) is called nonlinear Schrödinger equation. In the following this name is used for the function $g$ of a general type as well. In Sec. 1.3 the consequences of the symmetry of Eq. (1.12) are discussed. In Sec. 1.4 solutions of this equation are considered of the form of a nearly plane wave. From the point of view of image transformation in amplifiers, the most interesting are the solutions of the form of a complex structure beam. Nonlinear transformation of this structure can be described statistically. Approximate solutions with definite statistical properties are analyzed in Chs. 2, 3 for Eq. (1.12) (the case of a single paraxial beam with random modulation) and in Ch. 4 for system (1.8) (the case of two colliding wide paraxial beams).

## 1.3. Nonlinear Schrödinger equation

In this section several general properties of solutions of paraxial diffraction equation are examined. Method of moments [3, 8, 11] is briefly described. The method provides an averaged description of interaction between optical beams with simple structures and a nonlinear medium. Waveguide beam collapse in a self-focusing medium with a weak linear amplification is analyzed to illustrate the method.

Many properties of solutions of the paraxial diffraction equation allow an obvious physical interpretation. It is based on a similarity of Eq. (1.12) and equation of a dynamic system evolution. Longitudinal spatial coordinate is an analogue of time (see, for example, [12]). For a transparent medium (purely refractive nonlinearity) this system is conservative, and Eq. (1.12) may be derived from the principle of stationary action, intriducing Lagrangian

$$L = E^{*}E_{,0}\frac{i}{2} - E^{*}_{,0}E\frac{i}{2} - E^{*}_{,j}E_{,j}\frac{1}{2k} - U(EE^{*}), \tag{1.13}$$

with the potential

$$U(EE^{*}) = -\int_{0}^{EE^{*}} \operatorname{Im} g(I)\,dI. \tag{1.14}$$

Here and below, subscript after comma denotes derivative with respect to spatial coordinates. Latin letters take the values of 1 and 2, Greek letters take the values of 0, 1, and 2, with the conditions of

$x_1 = x$ , $x_2 = y$ , $x_0 = z$ (summation is performed over repetitive indices). In [18] this Lagrangian is written in a somewhat less symmetric (but equivalent) form. For simplicity, we restrict ourselves to the case of two spatial coordinates (as the most physically obvious) and unidirectional propagation of radiation, since inclusion of a counter-propagating beam into the same Lagrangian would have made the analogy with a dynamic system less obvious. We note that $U(I) = -\gamma I^2/2$ corresponds to a third-order nonlinearity and $U(I) = -\mathrm{Im}(g_0)\ln(1 + I/I_s)$ corresponds to a saturation nonlinearity with a purely imaginary increment $g$.

Requirement of extremum action $S = \int L d^3x$ gives Lagrange-Euler equations in the form of

$$(\partial L/\partial E_{,\alpha})_{,\alpha} = \partial L/\partial E. \tag{1.15}$$

Substitution of (1.13) into (1.15) gives diffraction equation (1.12).

It is convenient sometimes to examine the properties of the solutions of Eq.(1.12) (or (1.15)) proceeding from the principle of stationary action rather than the equations themselves. In particular, action invariance with respect to continuous transformations of the field $E$ and the coordinates $x$, according to the Noether theorem [16,17,85], results in conservation of the corresponding number of integral quantities. They characterize such beam parameters as power, divergence, position, direction, size, and twisting. In [12,18] these integrals (not only in the sence of integrating over transverse coordinates but also as the integrals of motion of Eq.(1.15)) were used, in particular, for investigation of the beam stability with respect to collapse. These integral quantities will be called the Noether integrals below, even if they are not integrals of motion. Distortions of the conservation laws due to the medium activity are determined by derivatives of the Noether integrals with respect to the longitudinal coordinate. Equalities are valid of the form of

$$\Theta_{\alpha,\alpha} = 2\bar{g}\mathcal{R}, \tag{1.16}$$

where $\Theta_\alpha$ is the Noether current and $\mathcal{R}$ is readily expressed through the longitudinal components of the Noether currents. Concrete expressions for $\mathcal{R}$ and the zero component of the Noether current $\Theta_0$ are summarized in Table 1.

The Noether integrals are the integrals of the Noether currents.

$$\tilde{\Theta} = \int \Theta_0 \, d^2x. \tag{1.17}$$

According to the Gauss theorem, derivatives of the Noether integrals can be expressed as

$$\frac{d}{dz}\int \Theta_0 \, d^2x = \int 2\bar{g}\mathcal{R} d^2x. \tag{1.18}$$

Let us discuss the physical sence of the Noether integrals defined in Table 1. The integral of intensity $I$ corresponds to the total beam power. For intensity, Table 1 indicates only that if $g$ is a local field increment then $2g$ is a local intensity increment. Momentum $P_j$, moment $M$, "Erenfestian" $F_j$ (a natural name for the current corresponding to Erenfest theorem concerning the motion of the centre of mass) and Hamiltonian $H$ characterize beam tilt with respect to the optical axis ( $z$ axis, i.e. $X_0$ ), its twisting relative to this axis, its position relative to this axis and its divergence, respectively. In homogeneous and purely refractive media, conservation of integrals of

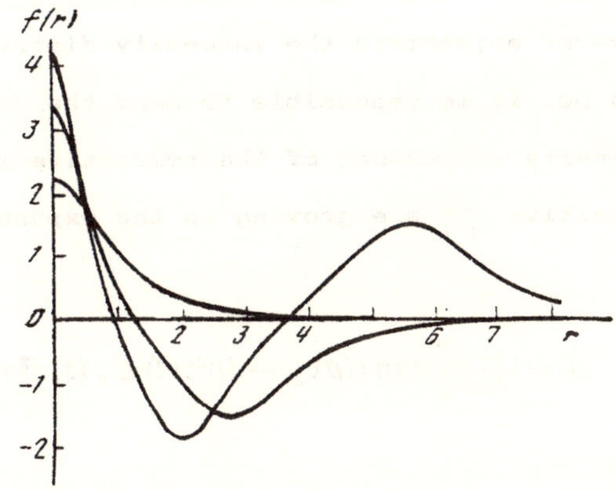


Fig. 1.1. Several solutions of Eq. (1.23) with boundary conditions $f'(0) = 0$, $f(\infty) = 0$. Waveguide configuration of the form of (1.22) corresponds to each of them.

Table 1. Quantities determining distortions of the Noether conservation laws at Re$g \neq 0$. 1-notes

| $\Theta_0$ | $\mathcal{R}$ | 1 |
|---|---|---|
| $I = EE^*$ | $I$ | — |
| $P_j = iEE^*_{,j}/2 - iE_{,j}E^*/2$ | $P_j$ | $j = 1, 2$ |
| $M = P_1x_2 - P_2x_1$ | $M$ | — |
| $F_j = P_jz - kx_jI$ | $F_j$ | $j = 1, 2$ |
| $H = E_{,j}E_{,j}/(2k) + U(I)$ | $H + N$ | $N = U'I - U - I_{,j,j}/(2k)$ |
| $D = P_jx_j - 2Hz$ | $D - Nz$ | $U(I) = -\gamma I^2/2$, $\gamma = \text{const}$ |
| $Q = Ikx^2/2 - Dz - Hz^2/2$ | $Q + Nz^2$ | $U(I) = -\gamma I^2/2$, $\gamma = \text{const}$, $x^2 = x_jx_j$ |

Table 2. Approximate values of $f(0)$ and integrals of the form of (1.25) for several solutions of Eq. (1.22). 1 - number of nodes.

| 1 | f (0) | $\tilde{U} = \tilde{I} = \tilde{C}$ | $\tilde{Q}$ |
|---|---|---|---|
| 0 | 2,2062 | 1,8262 | 1,106 |
| 1 | 3,332 | 12,28 | 54,0 |
| 2 | 4,15 | 31,17 | 395 |
| 3 | 4,83 | 58,5 | 1465 |

these currents limits the variety of nonlinear beam transformations. In particular, however asymmetric the intensity distribution at the medium input could be, it is impossible to bend the beam as a whole utilizing the intensity dependence of the refractive index. If the components with positive $p_j$ are growing in the expansion in plane waves:

$$E(x_\perp, z) = \frac{1}{2\pi}\int \exp(ipx_\perp - ip^2 z/2)\, E_p(z)\, d^2p \tag{1.19}$$

then the components with negative $p_j$ are growing too, since the total momentum

$$\tilde{P}_j = \int P_j d^2x = \int_D p_j E_p E_p^* \, d^2p \tag{1.20}$$

does not vary. We note that the modulus squared of the Fourier component of the field $E_p$ describes the far-field intensity distribution, so it has a straightforward physical interpretation.

Hamilton conservation means that if the potential $\mathcal{U}$ is decreasing on the average (for instance, owing to the beam compression), then the components $E_p$ with with large values of $p^2 = p_j p_j$ are growing.

Conservation of the dilatation current $\mathcal{D}$ and the inversion current $\mathcal{Q}$ gives the law of the beam effective cross section variation in a medium with a third-order nonlinearity: $\langle x^2 \rangle = \int I x^2 d^2x / \int I d^2x$. The reference point is chosen here in such a way that $\langle x \rangle = 0$. This area may depend on the longitudinal coordinate as a second-order polynom only: $\langle x^2 \rangle \tilde{I} = \tilde{Q} + \tilde{\mathcal{D}} z + \tilde{H} z^2$.

The presence of medium activity leads to variation of the integrals of the form of (1.17). They seize to be integrals of motion. Moreover, the currents expressed by quadratic functions of the field

($I, P_j, M, F_j$) are amplified with one and the same increment. In the case of linear amplification, the corresponding integrals grow exponentially with identical rates. For more complex currents, additional term appears in the expression for $\mathcal{R}$.

Formulae of the form of (1.18) are direct consequences of symmetry of the paraxial diffraction equation. Table 1 is used in the following to control the adequacy of approximate solutions of Eq.(1.12). If the field has a simple transverse structure, Table 1 can give information about the beam behaviour in an amplifying medium with nonlinear refraction. An example of such an estimate is presented below.

Let us consider localized centrosymmetric steady-state (waveguide) solutions of Eq.(1.12) for the case of a purely refractive medium with a third-order nonlinearity ($\mathcal{U}(I) = -\gamma I^2/2$, $\bar{g} = 0$). In a cylindric frame ($X_1 = \rho \cos\alpha$, $X_2 = \rho \sin\alpha$), Eq.(1.12) has the form of

$$i \frac{\partial E}{\partial z} + \left(\frac{1}{\rho} \frac{\partial}{\partial \rho} \rho \frac{\partial}{\partial \rho} + \frac{1}{\rho^2} \frac{\partial^2}{\partial \alpha^2}\right) E + \gamma E E^* E = 0. \qquad (1.21)$$

In [14,15] solutions of Eq.(1.21) are described of the following form:

$$E = \sqrt{\nu/\gamma}\, f\left(\sqrt{2k\nu}\, \rho\right) \exp(i\nu z), \qquad (1.22)$$

where $\nu$ = const, $f(r)$ is a real function satisfying the equation

$$f'' + f'/r - f + f^3 = 0. \qquad (1.23)$$

Requirement of the solution smoothness at the zero point results in the boundary condition $f'(0) = 0$. Localization of the solution implies that the function decays for large $r$ so the condition $f(\infty) = 0$ appears. Solution of Eq.(1.22) satisfying these boundary conditions corresponds to a confined beam with the moment $\tilde{M} = 0$. Several solutions of the kind are reproduced in Fig. 1.1 for illustration. We note that the first of them (without nodes) is presented in [13]. Instability of such solutions was discussed in [18].

From symmetry and constancy (along $z$ axis) of the solutions of the form of (1.22), it follows that $\tilde{P}_j = \tilde{F}_j = \tilde{M} = \tilde{H} = \tilde{\mathcal{D}} = 0$ for them. Nonzero integrals may be presented in a convenient form

$$\tilde{I} = \int EE^* \, d^2x = \frac{\pi}{k\gamma} \breve{I}, \qquad \tilde{Q} = \int EE^* \frac{x^2}{2} \, d^2x = \frac{\pi}{2k\gamma\nu} \breve{Q},$$
$$\tilde{U} = \int U \, d^2x = -\frac{\pi\nu}{k\gamma} \breve{U}, \qquad \tilde{C} = \int \frac{1}{2k} E_{,j} E^*_{,j} \, d^2x = \frac{\pi\nu}{k\gamma} \breve{C}, \tag{1.24}$$

where

$$\breve{I} = \int_0^\infty (f(r))^2 r \, dr, \qquad \breve{Q} = \int_0^\infty (f(r))^2 (r^2/2) r \, dr,$$
$$\breve{U} = \int_0^\infty ((f(r))^4/2) r \, dr, \qquad \breve{C} = \int_0^\infty (f'(r))^2 r \, dr. \tag{1.25}$$

From the condition $\tilde{H} = 0$ we obtain $\breve{U} = \breve{C}$. It follows from this equality, Eq. (1.23) and the boundary conditions that $\breve{U} = \breve{I}$.

Approximate magnitudes of $f(0)$, $\breve{I}$ and $\breve{Q}$ for several functions are presented in Table 2.

Possible dimensions and intensities of the beams determined by the solutions of the form of (1.22) are estimated in [15]. For instance, at $k \approx 10^5$ /cm, $\gamma = kn_2/n_0 \approx 10^{-10}$ (cm·W)$^{-1}$, self-focusing power is $\tilde{I} \approx \pi/(k\gamma) \approx 10^6$ W. For intensities of the order of $10^8$ W/cm, typical cross section of a waveguide beam is $\langle x^2 \rangle \approx 10^{-2}$ cm$^2$, typical scale length of the configurations shown in Fig. 1.1 is about 1 mm (the corresponding value of the parameter $\nu$ is about $10^{-2}$ $cm^{-1}$.

To illustrate the application of Table 1 we consider transformation of a beam spatial structure in a medium with third-order self-focising nonlinearity and linear amplification. It follows from Table 1 that

$$\dot{\tilde{I}}(z) = 2\bar{g}\tilde{I}, \qquad \dot{\tilde{H}}(z) = 2\bar{g}\tilde{H} + 2\bar{g}\tilde{U},$$
$$\dot{\tilde{D}}(z) = 2\bar{g}\tilde{D} - 2\bar{g}\tilde{U}z, \qquad \dot{\tilde{Q}}(z) = 2\bar{g}\tilde{Q} + 2\bar{g}\tilde{U}z^2, \tag{1.26}$$

where a dot denotes derivation $d/dz$; $\bar{g}$ = Re $g$ = const. At $\bar{g} \neq 0$ system (1.26) is not closed. $\tilde{U}(z)$ should be expressed basing on additional considerations. We suggest that the intensity distributions in different cross sections are similar. Then $\langle x^2\rangle \tilde{U}/\tilde{I}^2$ = const. We present the beam effective cross section in the form of

$$\langle x^2\rangle = 2\,(\tilde{Q} + \tilde{D}z + \tilde{H}z^2)/(\tilde{I}k). \tag{1.27}$$

From this equation the potential is expressed as

$$\tilde{U}(z) = \tilde{U}(0)\,\tilde{Q}(0)\,\tilde{I}^2/((\tilde{Q} + \tilde{D}z + \tilde{H}z^2)\,\tilde{I}^2(0)). \tag{1.28}$$

This formula enables one to try the solution of system (1.26) in the form of

$$\tilde{I}(z) = \tilde{I}(0)\exp(2\bar{g}z), \qquad \tilde{H}(z) = h(z)\exp(2\bar{g}z),$$
$$\tilde{D}(z) = d(z)\exp(2\bar{g}z), \qquad \tilde{Q}(z) = q(z)\exp(2\bar{g}z). \tag{1.29}$$

From (1.25), (1.22), and (1.26) we obtain a closed system

$$\dot{q} = \dot{h}z^2, \qquad \dot{d} = \dot{h}z, \qquad \dot{h} = 2\bar{g}U(0)\,q\exp(2\bar{g}z)/(q + zd + z^2h). \tag{1.30}$$

System (1.30) describes evolution of the principal beam parameters in the process of propagation in a medium. It is a particular case of equations of the moment method [8]. Let us examine the case of small $\bar{g}$. In the zero order of the perturbation theory we have

$$q = \tilde{Q}(0) + O(\bar{g}), \quad d = \tilde{D}(0) + O(\bar{g}), \quad h = \tilde{H}(0) + O(\bar{g}). \tag{1.31}$$

From formulae (1.31) and (1.27) it follows that in the absence of amplification and at $\tilde{H}(0) < 0$, the beam contracts at the distance of $z \approx L_f = ((d^2 - 4hd)^{1/2} - d)/2$ from the cross section $z = 0$ where the boundary condition is imposed. The intensity grows thus much that Eq. (1.12) becomes inapplicable. This case is considered in [8-11] in a greater detail. The length $L_f$ is called the Kelley length [86].

We turn now to the first order of the perturbation theory. Substituting (1.31) into (1.30) we get

$$\begin{aligned} q &= \tilde{Q}(0) + 2\bar{g}\tilde{U}(0)z^3/3 + O(z^4), \\ d &= \tilde{D}(0) + 2\bar{g}\tilde{U}(0)z^2/2 + O(z^3), \\ h &= \tilde{H}(0) + 2\bar{g}\tilde{U}(0)z + O(z^2). \end{aligned} \tag{1.32}$$

For initial distributions of the form of (1.22) when $\tilde{D}(0) = \tilde{H}(0) = 0$ an estimate for the Kelley length follows from (1.32)

$$L_f \sim \left(\frac{5Q(0)}{3\bar{g}U(0)}\right)^{1/3} = \left(\frac{5\breve{Q}}{3\bar{g}\nu\breve{U}}\right)^{1/3}. \tag{1.33}$$

It can be seen from (1.31) that for this case $O(\bar{g}^m)$ can be substituted for $O(z^m)$ in formulae (1.32) at $m = 2$, 3, and 4. In particular, at $\nu = 100\ \mathrm{cm}^{-1}$ and $\bar{g} = 5\cdot 10^{-3}\ \mathrm{cm}^{-1}$ formula (1.33) gives $L_f \approx 1$ m. We note that for $gL_f \approx 1$, estimation (1.33) only gives the order of magnitude, since formula (1.28) is based on the assumption about the similarity of intensity distributions in different cross sections, which becomes invalid.

Estimation (1.33) was made for the case, when amplification is not saturated, and the increment is supposed to be uniform across the beam. Let us examine how the spatial nonuniformity of the increment prevents the beam from collapsing. We express the rate of the variation of the beam effective cross section, according to Table 1

$$d\,\langle x^2\rangle/dz = \mathfrak{R} + \mathfrak{A}. \tag{1.34}$$

$\mathfrak{R}$ denotes the term associated with refraction:

$$\mathfrak{R} = \frac{\int D + 2Hz\, d^2x}{(k/2)\int I\, d^2x}, \tag{1.34a}$$

and $\mathfrak{A}$ denotes the term associated with amplification:

$$\mathfrak{A} = \frac{\int 2\bar{g} I x\, d^2x}{\int I\, d^2x} - \frac{\int I x^2\, d^2x}{\int I\, d^2x}\,\frac{\int 2\bar{g} I\, d^2x}{\int I\, d^2x}.$$

We rewrite this formula in the form of

$$\mathfrak{A} = -2\left(\langle \bar{g}x^2\rangle_I - \langle \bar{g}\rangle_I\,\langle x^2\rangle_I\right). \tag{1.34b}$$

where $\langle\ \rangle_I$ denotes averaging with the weight of $I/\int I d^2x$ . We note that if the transverse intensity distribution is bell-shaped and the function $\bar{g}(I)$ decreases with the intensity growing, then correlation $\langle \bar{g}x^2\rangle_I - \langle \bar{g}\rangle_I \langle x^2\rangle_I$ is negative. Thus, formulae (1.34) and (1.34b) describe defocusing action of amplification. It was discussed, in particular, in [5-11,87].

The nonlinear wave equation of diffraction permits a paraxial monochromatic beam in a homogeneous nonlinear medium to be taken as a dynamic system (one of spatial coordinates plays the role of time). Just as in the case of other dynamic systems, it is helpful to use the principle of steady-state action, along with equations of motion. For purely refractive nonlinearity, this system is conservative and a symmetry with respect to continuous transformations results in conservation of certain currents. Distortions of these conservation laws in the presence of amplification are described by Table 1. These laws describe the influence of amplification on the power, divergence, position, direction and cross section of the beam. Quantitative consideration of the violations of conservation provides the way to check up the adequacy of approximate (in particular, numerical) solutions of Eq.(1.12). An example of such a checkup is given below in Sec.2.6. With certain assumptions concerning the beam structure, the equations of the form of (1.18) may give a closed system. An example of such a system for the case of a field with a complex structure will be given in Sec.2.2. Still one more example is a system (1.30) written in the approximation of a beam with a constant shape. This system describes the influence of amplification on the waveguide solution instability.

A question may arise of other possible amplification effects on the self-focusing instability. The case of a nearly plane wave is important from the point of view of practical applications. Since a plane wave with small perturbations is a particular case of the partially modulated field studied in the following chapters, consideration of the influence of the amplification saturation on the self-focusing instability is interesting also as an illustration. The next section is devoted to this case.

## 1.4. Linearized diffraction equation

In the following chapters propagation of partially modulated radiation consisting of a plane wave and a component completely modulated in a cross section is considered in a nonlinear medium. In this section, a limiting case of a weak modulation (a weak perturbation on the background of an intense plane wave) is examined as a preliminary analysis.

Let us write Eq. (1.12) for the case of a third-order refractive nonlinearity and an active nonlinearity of a general type in the form of

$$\left(\frac{\partial}{\partial z} + \frac{\Delta_\perp}{2ik}\right) E = g\,(EE^*)\,E = i\gamma EE^*E + \bar{g}\,(EE^*)\,E, \tag{1.35}$$

where $\mathrm{Im}\,\bar{g}(I) = 0$, $\gamma$ = const is a nonlinear refraction constant. Extracting a plane wave, we shall try an approximate solution of Eq. (1.35) in the form of

$$E = A\,(z)\,(1 + \mathscr{B}\,(z)\cos\,(px + v)), \quad |\,\mathscr{B}\,| \ll 1, \quad v = \text{const}. \tag{1.36}$$

In linear approximation in $\mathscr{B}$, substitution of (1.36) into (1.35) gives a system

$$dA/dz = i\gamma A + \bar{g}A, \tag{1.37}$$

$$d\mathscr{B}/dz = (i\,(\gamma a - q) + \bar{g}'a)\,\mathscr{B} + (i\gamma a + \bar{g}'a)\,\mathscr{B}^*, \tag{1.38}$$

where $a = AA^*$, $\bar{g} = \bar{g}(a)$, $\bar{g}' = d\bar{g}/da = \bar{g}'(a)$, $q = p^2/(2k)$. It is convenient to assume $\mathscr{B} = \mathscr{M} + i\mathscr{N}$ and to put down separate equations for the real part and for the imaginary part

$$d\mathscr{M}/dz = q\mathscr{N} + 2\bar{g}a\mathscr{M}, \quad d\mathscr{N}/dz = (2\gamma a - q)\,\mathscr{M}. \tag{1.39}$$

Eliminating $\mathcal{M}$ from the system, we obtain the second-order equation

$$d^2\mathcal{N}/dz^2 - (2\bar{g}'a + 4a\bar{g}/(2\gamma a - q))\,d\mathcal{N}/dz - (2\gamma a - q)\,q\mathcal{N} = 0. \quad (1.40)$$

At $q < 2\gamma a$ and weak amplification ($q \ll \gamma a$, $\bar{g}' \ll \gamma$) this equation has a rapidly growing solution, moreover, at $\bar{g} = 0$ the growth is exponential, and the increment is

$$\varkappa = \sqrt{(2\gamma a - q)\,q}\,. \quad (1.41)$$

Let us consider Eq.(1.40) as the equation of motion of a material point with unit mass in the potential $-\varkappa^2\mathcal{N}/2$. For sufficiently rapid decay of the function $\bar{g}$, the coefficient multiplying the second term of formula (1.40) gives quasifriction. We will show that the saturation nonlinearity does not give such a noticeable quasifriction that would affect the beam stability radically.

Let us analyze perturbations with $q = \gamma a$, that grow particularly rapidly in the absence of amplification. For them Eq.(1.40) takes the form of

$$d^2\mathcal{N}/dz^2 - 2\,(\bar{g}'a + 2\bar{g})\,d\mathcal{N}/dz - (\gamma a)^2\,\mathcal{N} = 0. \quad (1.42)$$

If $\bar{g}(a) = g_0 I_s/(I_s + a)$, then $-2(\bar{g}'a + 2\bar{g}) = -2g_0 I_s a/(I_s + a)^2 < 0$. So, quasifriction turned out to be negative. It means that inhomogeneity smoothing due to larger increment in the regions with lower intensity cannot stabilize a plane wave at any $g_0$ and $I_s$. Stabilization could occur at a more rapid decrease of the increment with intensity growing, for instance, due to multiphoton absorption. In [88] nonlinear response of neodimium laser glasses was investigated. Although two-photon absorption was observed in these media, nonlinearity associated with this mechanism was shown to be at least an order of magnitude lower than the refractive nonlinearity. In this connection, it looks more promising to utilize resonant refractive nonlinearity proposed in [89].

Let us find out, how essential the influence of saturable amplification on the development of the self-focusing instability of a plane wave can be. Since Eq. (1.42) is written for $q = \gamma a$, it does not describe evolution of a perturbation with a fixed value of $q$. Amplification of a plane wave makes the perturbations with decreasing $q$ values the most unstable ones. For estimation of a perturbation gain coefficient, the so-called decay integral is usually used in this case (see, for example, [40,41]):

$$\Phi = \int_0^\infty \gamma a\,(z)\,dz, \qquad \Lambda = \left| \frac{\mathscr{B}\,(z)}{\mathscr{B}\,(0)} \right|^2 \approx \exp\,(2\Phi). \tag{1.43}$$

Due to the mechanism just considered (increment decrease in the regions with higher intensities and variations of a transverse wavenumber corresponding to the most unstable mode), expression (1.43) gives an overestimation. Fig. 1.2 shows amplification factors for a perturbation with a fixed value of $\tilde{q} = q/(\gamma I_S)$ at $g_0 = 0.2\,\gamma I_S$ (a); $\gamma I_S$ (b); $5\gamma I_S$ (c). Since the growth of every component depends on its initial phase, it is natural to follow the growth of the maximum (over the initial values of Arg $\mathscr{B}(0)$)[1] coefficient $\Lambda_q$. Just this quantity is shown in the lower graphs for several values of $\tilde{q}$. Estimates from the decay integrals (1.43) are snown by the dashed lines for comparison. For $\Phi \approx 1$, the perturbation growth is lower than the estimate $\exp(2\Phi)$ by a factor of about two in Fig. 1.2, a

---

1. Here and below, Arg denotes argument of a complex number reduced to the interval $(-\pi, \pi]$. Denotation arg is used if the interval is not fixed.

(weak amplification), by a factor of three in Fig 1.2,b (active nonlinearity of the same order), and by a factor of four in Fig.1.2,c (active nonlinearity is stronger than refractive one). Though, when the graphs are made in a logarithmic scale, the difference is not very large. So, we shoud conclude that the decay integral (1.43) can qualitatively describe the perturbation growth, in correspondence with common notion.

Note that linearized equation (1.38) describes not only the perturbation growth of a plane wave at the transverse wavenumber values $p^2/(2k) \approx \gamma a$ but also amplitude-phase distortions at $p^2/(2k) \gg \gamma a$. At large $p^2/(2k\gamma a)$ one can try approximate solution of Eq. (1.38) in the form of

$$\mathcal{B}(z) = \tilde{\mathcal{B}}(z) \exp\left(ip^2z/(2k) + i\int_0^z \gamma a(\zeta)\, d\zeta\right), \tag{1.44}$$

where $\tilde{\mathcal{B}}$ satisfies the equation

$$d\tilde{\mathcal{B}}/dz = \bar{g}' a\tilde{\mathcal{B}} + (i\gamma a + \bar{g}^{*}a)\,\tilde{\mathcal{B}} \exp\left(ip^2z/(2k) - i\int_0^z \gamma a(\zeta)\, d\zeta\right). \tag{1.45}$$

Neglecting the rapidly oscillating last term, we get approximate solution

$$\mathcal{B}(z) = \mathcal{B}(0) \exp\left(-ip^2/(2k) + i\int_0^z \gamma a(\zeta)\, d\zeta + i\int_0^z \bar{g}'(a(\zeta))\, d\zeta\right). \tag{1.46}$$

Noting that $\gamma = -\mathrm{Im}\,\bar{g}'$ we rewrite this formula in the form of

$$\mathcal{B}(z) = \mathcal{B}(0) \exp\left(-ip^2z/(2k) + \int_0^z g'(a(\zeta))\, a(\zeta)\, d\zeta\right). \tag{1.47}$$

In this approximation, nonlinear distortions of amplitude and phase of a short-wave perturbation do not depend on $p$, all the components are distorted in the same way. A weak field acquires, as a whole, a phase

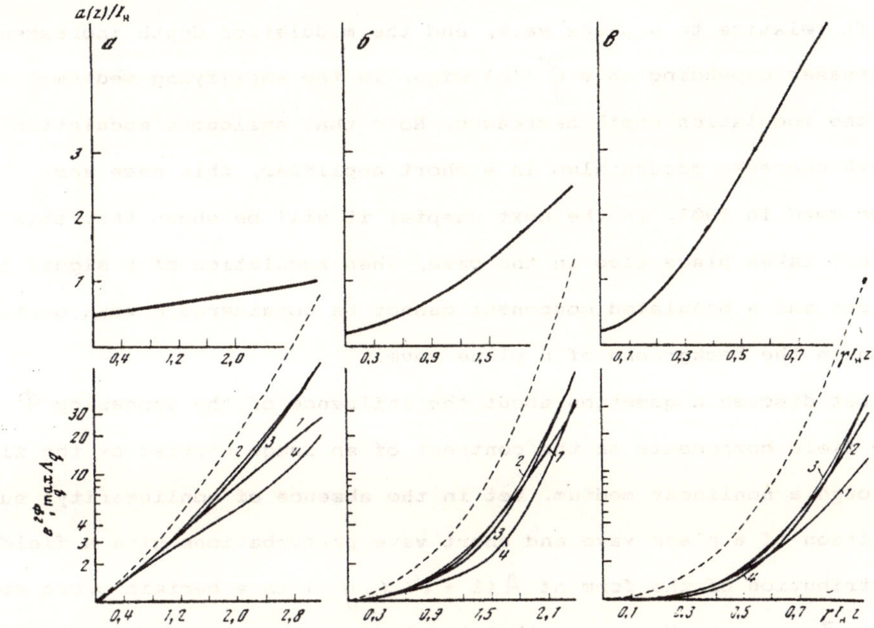


**Fig. 1.2. Dependence of normalized intensity $a(z)/I_s$ of the plane wave (solutions of Eq. (1.37); upper graphs) and of the amplification coefficients of perturbations $\max\Lambda_q$ (lower graphs) on dimensionless longitudinal coordinate $\gamma I_s z$ at various values of parameters**

**a – $g_0 = 0.2\gamma I_s$, (0) = $0.5 I_s$, $\tilde{q}$ = 0.1 (1), 0.5 (2), 1 (3), 1.5 (4)**

**b $g_0 = \gamma I_s$, $a(0) = 0.2 I_s$, $\tilde{q}$ = 0.5 (1), 1 (2), 1.5 (3), 2 (4)**

**c – $g_0 = 5\gamma I_s$, $a(0) = 0.2 I_s$, $\tilde{q}$ = 1 (1), 2 (2), 3 (3), 4 (4)**

**Dashed curves in the lower graphs correspond to an estimate $e^{2\phi}$ (1.43) using decay integrals.**

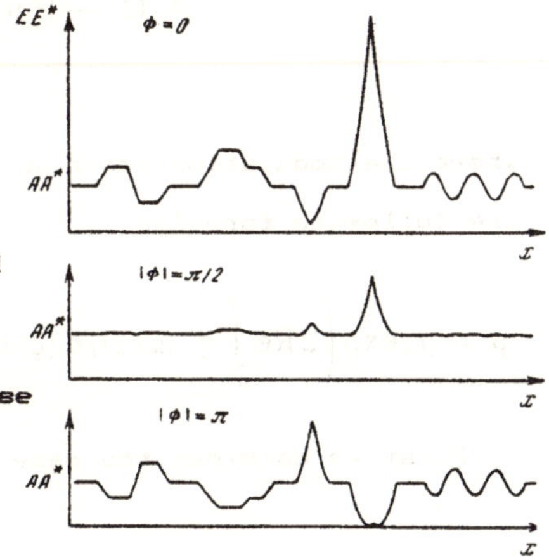


**Fig. 1.3. Variations of image features in the presence of dephasing: dependence of the intensity of the field $E = A\cdot(1 + U(X)\exp(i\phi))$, which is real at $\phi = 0$, on the transverse coordinate.**

shift relative to a plane wave, and the modulation depth increases or decreases depending on a $\bar{g}'(a)$ sign. In the amplifying medium $\bar{g}' < 0$, so the modulation depth decreases. Note that analogous modulation depth decrease occurs also in a short amplifier, this case was discussed in [90]. In the next chapter it will be shown that this effect takes place also in the case, when modulation of a signal is strong and a modulated component cannot be considered a weak perturbation on the background of a plane wave.

Let discuss a question about the influence of the dephasing $\Phi$ of the field components on the contrast of an image carried by the field through a nonlinear medium. Let in the absence of nonlinearity, superposition of a plane wave and short-wave perturbations give a field distribution of the form of $A(1+V(x,y))$ in a certain cross section $z$ = const. We assume that the typical magnitude of the function $V(x,y)$ modulus squared is a certain value $\mu_0$. It is natural to call this parameter the modulation depth of a field. Since this section deals with small perturbations on the background of an intense plane wave, $\mu_0 \ll 1$. Let the components of the field $V(x,y)$ with transverse wavenumber $p < (2k\gamma a)^{1/2}$ be absent. Then it follows from (1.47) that the field distribution in this cross section in the presence of nonlinearity is

$$A\cdot(1+\sqrt{\mu/\mu_0}\, V(x,\ y)\exp(i\Phi)), \tag{1.48}$$

where the modulation depth $\mu$ and the dephasing $\Phi$ are determined from the following formulae

$$\mu=\mu_0\exp\left(2\,\mathrm{Re}\int_0^z g'(a(\zeta))\,a(\zeta)\,d\zeta\right), \qquad \Phi=\mathrm{Im}\int_0^z g'(a(\zeta))\,a(\zeta)\,d\zeta. \tag{1.49}$$

First we consider the case when $V(x,y)$ is real. For instance, image of an amplitude transparency is focused on a plane $\zeta$ = const. Assuming that the intensity $I_{max} = a\left|1 + \sqrt{\mu}\cos\Phi\right|^2$ corresponds to the bright points of the initial field and $I_{min} = a\left|1 - \sqrt{\mu}\cos\Phi\right|^2$ corresponds to the dark points, we express the image contrast in the given cross section through the modulation depth

$$v = (I_{max} - I_{min})/(I_{max} + I_{min}) = 2\mu\cos(\Phi)/(1+\mu). \qquad (1.50)$$

Note that even at small modulation depths (for example, $\mu$ = 0.1), the image may be quite contrast (for example, $v$ = 0.6) if $\cos\Phi$ = 1.

To illustrate the dependence of the image contrast on the dephasing, let us consider a model example, in which a real (at $\Phi$ = 0) field in a given cross section depends on only one spatial coordinate (Fig.1.3). It is seen from the figure that at $|\Phi| = \pi/2$ the image almost vanishes, the contrast is equal to zero, and at $|\Phi| = \pi$ inversion (reversal) of the contrast occurs, i.e. initially bright spots become dark. In this sence, dephasing $|\Phi| = \pi$ results in formation of a negative image. Sharp dependence of contrast on dephasing can be used to measure nonlinear parameters of a medium. This question will be discussed in a greater detail in Sec.3.2 on the example of a purely refractive medium with a saturation type nonlinearity.

Note that in the case of an amplitude-phase image, when the real and the imaginary parts of $V(x, y)$ are independent, dephasing $\Phi \approx \pm\pi/2$ changes the appearence of an image radically, i.e. the amplitude modulation transforms into the phase one, and vice versa.

Thus, from the analysis of the linearized diffraction equation (system (1.37),(1.38)) it follows that in amplifying medium, the relative intensity of spatial Fourier components of a field with the diffraction length $L_d$ of the order of the nonlinear length $L_n$ grows slower than the estimation from the decay integral (1.43) predicts. Distortions of Fourier components with large values of the transverse wavenumber are described by two parameters, namely, the dephasing $\phi$ and the modulation depth $\mu$. Variations of these parameters are determined from formula (1.49). These distortions correspond to variations of a contrast of an image carried by the field. The image appearance depends on the dephasing stronger than on the modulation depth.

## 1.5. Interaction of colliding beams and smoothing action of amplification saturation (conclusions of Ch.1)

Interaction of colliding paraxial beams in a nonlinear medium is described by system (1.8). This system is used in Ch.4 to investigate interaction between colliding spatially modulated waves in an amplifier. In many cases the approximation is justified, in which the medium nonlinearity is characterized by a single complex function of intensity] namely, the local increment $g(I)$. The real part of the local increment corresponds to an ordinary gain coefficient, the imaginary part describes the refractive properties of the medium. In actual amplifiers, these magnitudes are often of the same order, so it is important to take into account both amplification saturation and refractive nonlinearity.

In the case of a single paraxial beam (the counter-propagating beam is absent), the first of Eqs.(1.8) transforms into nonlinear Schrödinger equation (1.12). Symmetry of this equation leads to certain integral relations for the solutions. They are determined from Table 1. If in a medium with a weak nonlinearity, a wide beam with small amplitude and phase fluctuations is propagating (a plane wave), the amplification saturation slows down the self-focusing instability growth to some extent in comparison with a widely used estimation from the decay integral (1.43). For small-scale perturbations of a plane wave at weak nonlinearity, Eq.(1.35) has approximate solution

(1.44) describing regular distortions of the field spatial structure. They can manifest themselves in contrast variations of the image carried by the field through the nonlinear medium. A more general case, when the intensity of the modulated component is comparable with the plane wave intensity, is analyzed in the next chapter.

# Chapter 2

## Interaction of a spatially modulated wave with complex structure and a plane wave: speckle-mode approximation

In this chapter, a wide beam with quasirandom spatial modulation is considered. The method utilized in Sec. 1.4 (linearization of the diffraction equation with respect to small perturbations of a plane wave) is generalized in the frameworks of the speckle-mode theory to the case of a finite modulation depth. Range of validity of the utilized method is estimated.

In Ch. 1 we mentioned the fact that certain integral characteristics were introduced in a number of papers concerning radiation propagation in a nonlinear medium. These characteristics are convenient to follow in calculations and give information about the beam variations in the medium. This approach is reasonable in the case, when the beam has a simple shape. However, for the field carrying a large amount of information, this description is insufficient. The field structure is too complicated to be characterized by several integral quantities. Introducing a large number of moments (this possibility is discussed in [9, 10]) we return, in fact, to numerical methods of solution of a partial differential equation. Meanwhile, as it is pointed out in the introduction, approximate description of the field distortions is of a great interest in this case. For such a description, certain assumptions about the field structure are needed. Let us concretize these assumptions.

Let us assume that a paraxial light beam is directed into a spatially uniform nonlinear medium, and this beam is modulated so that it can be considered a random one. Such a structure is typical of fields utilized, for example, in figurative or technical holography or in installations with phase conjugation. Let the basic parameters of such a field (for instance, modulation depth, speckle size, average intensity) be typical of the whole beam cross section. This situation occurs, in particular, in amplification of a complex image (for instance, a chip) containing a large number of more or less uniformly distributed details. Field distribution at the medium exit is determined by the solution of the nonlinear diffraction equation (1.12). The question arises, whether it is possible to say anything about the field distribution at the medium exit without exact solution of the nonlinear partial differential equation. For example, what the field distribution will be if one attempts to construct an image of an object that converted a plane wave into a field of complex structure with the help of an objective (Fig.2.1)? For small modulation depths, the answer to this question is given by linearization of Eq.(1.12) with respect to small perturbations of the plane wave (Sec.1.4).

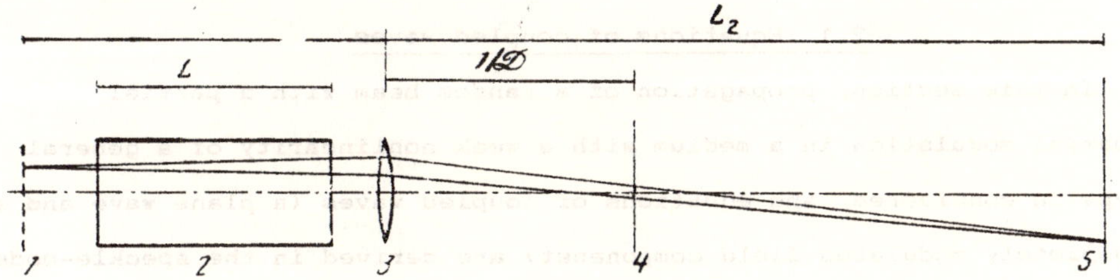


**Fig. 2.1. Layout of experiment on transformation of spatial structure of a complicaled field in a nonlinear medium**

**A plane wave illuminates a random transparency (1) from the left giving a field with complicated structure. This field passes through nonlinear medium (2). Lens (3) with an optical strength of $\mathcal{D} = 1/(L_1 + (n_0 - 1)L) + 1/L_2$ gives an image of the transparency on screen (5). Zernike filter (4) may be used to control the parameters of the output radiation.**

The purpose of this chapter is generalization of the perturbation method described in Sec. 1.4 to the case of a finite (not infinitesimal) modulation depth. Complexity of a wide beam spatial structure results in the fact that distribution of both amplitude and phase of the field across the beam varies due to diffraction to a greater extent than due to nonlinearity in the process of propagation. For a weak nonlinearity, a possibility appears to describe statistically its influence on the field structure. This approach allows changing from the limiting case of a weak modulation of a field to a more general case of a finite modulation depth.

## 2.1. Equations of coupled waves.

In this section, propagation of a random beam with a partial spatial modulation in a medium with a weak nonlinearity of a general type is considered, and equations of coupled waves (a plane wave and a completely modulated field componenet) are derived in the speckle-mode approximation.

It was shown in Ch.1 that a paraxial beam in a nonlinear medium with local response can be described by Eq.(1.12). Physical problem corresponds to determination of the field at the boundary of the nonlinear medium (for definiteness, in the cross section $z = 0$) as the boundary condition. Let $z$-axis be determined by the most intense spatial Fourier component of the field $E$. To follow the distortions of the field containing a plane wave and a completely modulated random component, let us try an approximate solution of Eq.(1.12) in the form of

$$\tilde{E}(x, y, z) = A(z) + B(z) f(x, y, z), \qquad (2.1)$$

where $f$ is a solution of the homogeneous equation $(\partial/\partial z + \Delta_\perp/(2ik))f = 0$. Note that in Sec.1.3, the symbol $f$ has different meaning. The field transverse distribution is described by a simple formula there. Without restriction of generality, we require that

$$\langle f \rangle_\perp = 0, \quad \langle ff^* \rangle_\perp = 1, \qquad (2.2)$$

where $\langle\ \rangle_\perp$ denotes averaging over transverse coordinates, the integration limits are determined by the beam size. Let $A(0) = \langle E(x,y,0) \rangle$. If we make transformation $E(x,y,z) \to E(x,y,z) \cdot |\langle E(x,y,0) \rangle_\perp| / \langle E(x,y,z) \rangle_\perp$, then the coefficients $A$ and $B$ can be considered nonnegative in the plane $z = 0$:

$$A\ (0) \geqslant 0, \quad B\ (0) \geqslant 0. \tag{2.3}$$

These conditions in combination with requirements (2.2) provide uniqueness of representation (2.1). We present exact solution of Eq. (1.12) in the form of

$$E = \tilde{E} + J, \tag{2.4}$$

where $J$ is a function orthogonal to both components of the initial field $E$. Since the intensity $\langle JJ^* \rangle_\perp$ growth restricts validity of the approximate solution of the form of (2.1) and distorts the image carried by the complex field $E$, the field $J$ will be called noise in the following.

Let us write down equations for the coefficients $A$, $B$ and the noise $J$

$$dA/dz = \langle g\,(EE^*)\,E \rangle_\perp, \quad dB/dz = \langle g\,(EE^*)\,Ef^* \rangle_\perp, \tag{2.5}$$

$$\left( \frac{\partial}{\partial z} + \frac{\Delta_\perp}{2ik} \right) J = R = g\,(EE^*)\,E - \langle g\,(EE^*)\,E \rangle_\perp - \langle g\,(EE^*)\,Ef^* \rangle_\perp f^*. \tag{2.6}$$

We note two circumstances concerning system (2.5),(2.6). First, it is exact in the sence that formulae (2.4),(2.1) give exact solution of Eq.(1.12). Second, Eqs.(2.5) provide minimum norm of the noise source $R$ in Eq.(2.6): its projections on the plane wave and on the completely modulated component $f$ of the field $\tilde{E} = A + Bf$ are subtracted from the right side of Eq.(1.12). In the papers dealing with phase conjugation [27-33] and holography [34-36], construction of an approximate solution of the field equations in the form of linear combination of solutions of linear homogeneous equations is termed the speckle-mode approximation, and equations for the coefficients in this linear combination are called equations of coupled waves. In this sence, Eqs. (2.5) are the equations of coupled waves.

It should be emphasized that the speckle-mode approximation implies a certain scale hierarchy

$$\lambda = 2\pi/k \ll l_\perp \ll L_d = kl_\perp^2 \ll L \ll L_\perp/\alpha = L_\perp k l_\perp, \tag{2.7}$$

where $\lambda$ is a wavelength, $k$ is a wavenumber, $l_\perp$ is a transverse modulation scale, $L_d$ is a diffraction length associated with this modulation, $L$ is a nonlinear medium length, $\alpha$ is a radiation angular divergence, $L_\perp$ is a beam width.

Note that in the conditions of [54], the last inequality was violated. In this paper, 500-fold amplification of brightness of an image of a complicated object (200 x 200 pixels) at the wavelength of $\lambda = 0.54\cdot10^{-4}$ cm is described in R6G dye-based active volume with dimensions of 10 mm × 1 mm × 1 mm. Absence of visible image distortions indicates that the transverse dimensions of the active region were underestimated in this paper.

The subsequent discussion applies to the range of $z$ values, where $BB^* \gg \langle JJ^* \rangle_\perp$, i.e. noise is small in comparison with the modulated component of the primary field. Assuming that the noise is small, we shall resolve Eqs. (2.5) using the substitution $E = \widetilde{E}$ .

Let us consider the question of boundary conditions of system (2.5), (2.6). We should set the field $E$ in the plane $z = 0$. The choice of $z$ axis is determined by the wavevector of the plane wave, if this component is present in the beam. There is a certain degree of freedom in denotation of the field component that varies in the transverse direction. One part of the component should be denoted by $Bf$ , and the other part by $J$ . On one hand, it is desirable to minimize the initial noise intensity. On the other hand, the integral calculations over the cross section in the right sides of system (2.5) are as cumbersome as the solution of(Eq. 1.12) itself is, so certain additional assumptions concerning the field $f$ structure are needed.

The spatial Fourier components of the field $E(X,y,0)$, for which the transverse wavenumber $p$ satisfies the condition $0 < |p| \lesssim \sqrt{k/L_n}$, are related to the noise $J$. Since the nonlinear length $L_n$ is large in comparison with the field diffraction length $L_d$ , the intensity of these Fourier components is low, i.e. $\langle JJ^* \rangle_\perp|_{z=0} \ll BB^*|_{z=0}$ .

It should be emphasized that if the field $E$ in the cross section $z = 0$ is a signal with a complicated structure, then we call a part of the signal the noise right away. This noise is relatively small for a sufficiently complicated signal. In an active medium, a small-scale noise components appear because of nonlinear self-action. These components occupy approximately the same spatial frequency range as the modulated component of the basic field does. In holography [32-34] analogous field appearing in a medium is called intramodulation noise. However it is expedient to use the term "noise" here, since the term amplifier noise is usually applied to a field created by spontaneous emission irrelevant of the transmission of a signal with a complicated structure through an amplifier.

So the choice of a function is determined by conditions (2.2),(2.3) and condition $\langle Jf^* \rangle_\perp = 0$. Conditions $\langle J \rangle_\perp|_{z=0} = 0$, $\langle Jf^* \rangle_\perp|_{z=0} = 0$ mean orthogonality of the noise $J$ and both components of the basic field $\tilde{E}$ in the plane $z = 0$. This orthogonality is preserved in all the cross sections, since the noise source $R$ in the right side of Eq. 2.6) is orthogonal to these components.

Note that in the case $A(0) = 0$, the first equation of system (2.5) results in $A(z) = 0$. This is the case of a completely modulated field. It was analyzed in [32,39], as we mentioned earlier. At $B(0) = 0$ the second equation of this system results in $B(z) = 0$. This is the case of a plane wabe which is not of a great interest. We imply below that $A(0) > 0$, $B(0) > 0$.

Let us examine system (2.5) neglecting the noise. Since the field $E$ corresponds to a random beam, we attribute the function $f$ to an ensemble of random functions. Assuming the transverse structure of the field $f$ to be sufficiently complicated and the average intensity, the modulation depth and the correlation radius to be typical of the whole beam cross section, we replace in (2.5) averaging over transverse coordinates $\langle \ \rangle_\perp$ by averaging over ensemble $\langle \ \rangle$. Distribution of values in each point of space is assumed a complex Gaussian one. Then Eqs. (2.5) change into a closed system.

$$dA/dz = X_a A, \quad X_a = \langle g(\widetilde{E}\widetilde{E}^*)\,\widetilde{E}\rangle / A,$$
$$dB/dz = X_b B, \quad X_b = \langle g(\widetilde{E}\widetilde{E}^*)\,\widetilde{E}\rangle / B, \tag{2.8}$$

where $\widetilde{E} = A + Bf$. The right sides in this system are expressed through double integrals over real and imaginary parts of the quantity $f$. We have changed to averaging over ensemble, and now if letter $f$ is encountered under the $\langle \ \rangle$ sign, it is a random quantity distributed with the density of $\exp(-ff^*)/\pi$ rather than a function of transverse coordinates. Parameters of this distribution are fixed by conditions (2.2).

With the help of change of integration variable $f \to f\cdot|B/A|\cdot A/B$ we present expressions for the increments $X_a$ and $X_b$ in the form of

$$X_a = X_a(a,\ b) = \langle g\,(a + 2\sqrt{ab}\,\mathrm{Re} f + bff^*)\,(1 + \sqrt{b/a}\,f)\rangle, \tag{2.9}$$
$$X_b = X_b(a,\ b) = \langle g\,(a + 2\sqrt{ab}\,\mathrm{Re} f + bff^*)\,(\sqrt{a/b} + f)\,f^*\rangle, \tag{2.10}$$

where $a = AA^*$ and $b = BB^*$ are the intensities of the plane wave and the modulated component of the field $E$. Note that the increments $X_a$

and $X_b$ are defined by these intensities and do not depend on the phases of arg( $A$ ) and arg( $B$ ), so the equations for intensities

$$da/dz = 2\mathrm{Re}\,(X_a)\,a, \tag{2.11a}$$
$$db/dz = 2\mathrm{Re}\,(X_b)\,b, \tag{2.11b}$$

described in [39] also form a closed system.

Thus, system (2.8) defines approximate solution (2.1) of Eq.(1.12). The uncertainty of this solution is discussed below in Sec.2.3. If one is interested only in intensities $a$ and $b$ of the field components, it is possible to use system (2.11). Note that system (2.11) can also be derived proceeding from the variation laws for the Noether integrals (see Sec.1.3), Eq.(1.12) being not used explicitly. This derivation is presented in the next section.

## 2.2. Derivation of system of equations for intensities of partially modulated field components on the basis of the results of Sec.1.3

In the previous section, system (2.11) describing variations of the intensities $a, b$ of the plane wave and the completely modulated field component in an extended amplifier was obtained on the basis of approximate solution of Eq.(1.12). In this section, alternative method for derivation is proposed, in which Eq.(1.12) is not used explicitly. It is based upon quantitative expressions for distortions of the Noether conservation laws found for a purely refractive medium. The idea of the derivation consists in expressing the derivatives of the random field component intensities $a, b$ using derivatives of the intensity and Hamiltonian mean values, in accordance with formula (1.26). In the course of the derivation, a new representation appears (formula (2.21)) for the increment $X_b$ of the modulated component of the basic field. This formula is needed for estimation of the noise intensity (see below, Sec.2.3).

We present quantitative expressions for distortions of the Noether conservation laws in the differential form: $I_{\mu,\mu} = 2gI$, $H_{\mu,\mu} = 2\bar{g}(H+N)$, where $\bar{g} = g(I)$, $\theta_{\mu,\mu} = \partial\theta_\mu/\partial x_\mu$, $\mu = 0,1,2$, $x_0 = z$, $x_1 = x$, $x_2 = y$. $\theta_\mu$ denote components of the Noether currents $\theta = (I, P_j, F_j, M, H, D, Q)$. In particular, $I_j = i(EE^*_{,j} - E_{,j}E^*)/(2k)$, $H_j = -(E^*_{,j}E_{,0} - E^*_{,0}E_{,j})/(2k)$.

Neglecting the noise, we suppose that the field has the form of (2.1). Using the variation laws for the Hamiltonian (see (1.16), Table 1), we find

$$\frac{dH}{dz} = \int 2\bar{g}\cdot(f_{,i}f^*_{,i}b/(2k) + U'I - I_{,i,i}/(4k))\, d^2x =$$
$$= \int (2\bar{g}bf_{,i}f^*_{,i}/(2k) + U'\,(I + H_{i,i}) + \bar{g}'I_{,i}I_{,i})\, d^2x,$$

where $\bar{g} = \mathrm{Re}\, g(I)$, $\bar{g}' = d\bar{g}/dI$, $I = EE^*$. Relation $2gI = I_{\mu,\mu}$ was used. Taking the averages and assuming statistical independence of the quantities $f$ and $f_{,j}$, we obtain

$$\frac{d}{dz}\langle H\rangle = \frac{d}{dz}\frac{1}{L_\perp}\int H\, d^2x = 2\,\mathrm{Re}\,\langle g + g'I\rangle\, b\left\langle \frac{1}{2k} f_{,i}f^*_{,i}\right\rangle + \left\langle U'\frac{\partial I}{\partial z}\right\rangle.$$

We present this expression in the form of

$$\frac{d}{dz}\langle H\rangle = 2\,\mathrm{Re}\,(X)\, b\left\langle \frac{1}{2k} f_{,i}f^*_{,i}\right\rangle + \frac{d}{dz}\langle U\rangle, \tag{2.12}$$

where

$$X = \langle g + g'I\rangle. \tag{2.13}$$

On the other hand, substituting the field of the form of (2.1) into the Hamiltonian definition (see Table 1) and differentiating, we get

$$\frac{d}{dz}\langle H\rangle = \frac{db}{dz}\left\langle \frac{1}{2k} f_{,i}f^*_{,i}\right\rangle + \frac{d}{dz}\langle U\rangle.$$

Comparing this formula with (2.12), we find the equation for the modulated component intensity

$$db/dz = 2\mathrm{Re}\,(X)\,b. \tag{2.14}$$

Note that from (2.14) and (2.11), equality $X = X_{\ell}$ follows. Subsequent derivations can be considered as a direct verification of this equality.

Let us prove that the increment $X$ defined by formula (2.13) coinsides with the increment $X_{\ell}$ of the modulated component of the basic field (formula (2.10)). Let us define the increment of a weak independent field by the formula

$$\tilde{X} = \lim_{J \to 0} \langle g\,(|\,\tilde{E} + J\,|^2)\,(\tilde{E} + J)\,J^* \rangle / \langle JJ^* \rangle. \tag{2.15}$$

When averaging, the weak field $J$ is considered to have distribution as random (i.e. Gaussian) as the field $f$.

First we show that $X = \tilde{X}$. For this purpose, we perform expansion in formula (2.15)

$$\tilde{X} = \lim_{J \to 0} \langle g \cdot (\tilde{E} + J)\,J^* + g' \cdot (\tilde{E}J^* + \tilde{E}^*J)\,\tilde{E}J^* + O\,(J^2) \rangle / \langle JJ^* \rangle.$$

Taking into account the independence of the quantities $\tilde{E}$ and $J$ and substituting the products of the avegages for the averages of their products, we get

$$\tilde{X} = \lim_{J \to 0} \langle gJJ^* + g'\tilde{E}\tilde{E}^*JJ^* \rangle / \langle JJ^* \rangle = \langle g + g'EE^* \rangle = X. \tag{2.16}$$

So formulae (2.10) and (2.15) define one and the same increment of a weak independent field.

Now we prove that $\tilde{X} = X_{\ell}$. Let us present the weak field in (2.15) in the form of $J = Gj$, where $\langle jj^* \rangle = 1$, $\langle j \rangle = 0$:

$$\widetilde{X} = \lim_{G\to 0} \langle g\,(|\,A + Bf + Gj\,|^2)\,(A + Bf + Gj)\,j^*G^*\rangle/\langle GG^*\rangle. \tag{2.17}$$

Here averaging is performed over $f$ and $j$ in assumption that these quantities are independent. Changing over to new integration variables

$$h = fB/\widetilde{B} + jG/\widetilde{B}, \quad s = -fG/\widetilde{B} + jB/\widetilde{B}; \quad \widetilde{B} = (b + GG^*)^{1/2}, \tag{2.18}$$

we reduce (2.15) to

$$\widetilde{X} = \lim_{G\to 0} \langle g\,(|\,A + \widetilde{B}h\,|^2)\,(A + \widetilde{B}h)\,(Gh^*/\widetilde{B} + B^*s^*/\widetilde{B})/G\rangle. \tag{2.19}$$

Performing averaging over $S$ in (2.19), we obtain

$$\widetilde{X} = \langle g\,(|\,A + Bh\,|^2)\,(A + Bh)\,h^*/B\rangle = \langle g\,(|\,A + Bf\,|^2)\,(A + Bf)\,f^*\rangle/B \tag{2.20}$$

Comparing this relation with (2.10), (2.16), and (2.13), we get

$$\widetilde{X} = X_b = \langle g\widetilde{E}f^*\rangle/B = X = \langle g + g'\widetilde{E}\widetilde{E}^*\rangle. \tag{2.21}$$

Eq.(2.11b) follows from this relation and Eq.(2.14).

To obtain (2.11a), we express the avegage intensity in the form of

$$\langle I\rangle = L_\perp^{-2} \int \widetilde{E}\widetilde{E}^*\, d^2x = a + b, \tag{2.22}$$

and the average value of its derivative from Table 1 is presented in the form of

$$d\,\langle I\rangle/dz = da/dz + db/dz = 2\,\langle \mathrm{Re}\, gI\rangle. \tag{2.23}$$

Taking into account the definitions of the effective increments in formulae (2.8) (but not using the formulae themselves), we express the right side of (2.23) through these increments

$$\mathrm{Re}\,\langle 2gI\rangle = 2\,\mathrm{Re}\,(a\,\langle g\tilde{E}\rangle/A + b\,\langle g\tilde{E}f^{*}\rangle/B) = 2\,\mathrm{Re}\,(X_a)\,a + 2\,\mathrm{Re}\,(X_b)\,b. \quad (2.24)$$

From comparison of (2.23) and (2.24) we get

$$dA/dz + db/dz = 2\,\mathrm{Re}\,(X_a)\,a + 2\,\mathrm{Re}\,(X_b)\,b. \quad (2.25)$$

Eq.(2.11a) follows from (2.25) and (2.11b).

So, system (2.11) is a direct consequence of assumptions about the field $E$ structure and the variation laws for the Noether integrals (intensity and Hamiltonian) found in Sec.1.3. Formula (2.21) is used in the next section in investigations of relative intensities of those nonlinear field distortions that cannot be described by approximate solution (2.1).

## 2.3. Estimation of noise intensity

In Sec.1.1, approximate solution (2.1) of the paraxial diffraction equation (1.12) was constructed. Difference $J$ between the exact solution $E$ and the approximate solution $\tilde{E}$ was called the noise. In the present section, the order of magnitude of its intensity $\langle JJ^{*}\rangle$ is estimated.

As it was noted earlier, The noise is orthogonal to both components of the basic field $\tilde{E}$. It follows from the condition $\langle J\rangle|_{z=0} = 0$ and orthogonality of the right side of Eq.(2.6) and these components. Let us introduce denotations $\tilde{g} = g(\tilde{E}\tilde{E}^{*})$ and $\tilde{g}' = g'(\tilde{E}\tilde{E}^{*})$. The noise source is written in the form of

$$R = gE - \langle gE\rangle_{\perp} - \langle gEf^{*}\rangle_{\perp} f. \quad (2.26)$$

In the linear approximation in $J$ we have

$$gE = \widetilde{g}\widetilde{E} + (\widetilde{g} + \widetilde{g}'\widetilde{E}\widetilde{E}^*)\, J + \widetilde{g}'\widetilde{E}^2 J^*. \tag{2.27}$$

Substituting (2.27) into (2.26), we get

$$R = \widetilde{g}\widetilde{E} - X_a A - X_b B + (\widetilde{g} + \widetilde{g}'\widetilde{E}\widetilde{E}^*)\, J + \widetilde{g}'\widetilde{E}^2 J^* - \langle(\widetilde{g} + \widetilde{g}'\widetilde{E}^*\widetilde{E})\, J\rangle_\perp - \\ - \langle(\widetilde{g} + \widetilde{g}'\widetilde{E}^*\widetilde{E})\, f^* J\rangle_\perp f - \langle\widetilde{g}'\widetilde{E}^2 J^*\rangle_\perp - \langle\widetilde{g}'\widetilde{E}^2 J^* f^*\rangle_\perp f + O(J^2). \tag{2.28}$$

Assuming that the noise phase does not correlate with the field $f$ phase, we neglect averaging in the right side of (2.28). Assuming that the spatial structure of the fields $f$, $J$ is sufficiently complicated (so that transition to ensemble averaging < > is possible) and neglecting their mutual correlation, we replace the coefficients multiplying $J$ and $J^*$ by their average values, and substitute the result into (2.6):

$$\left(\frac{\partial}{\partial z} + \frac{\Delta_\perp}{2ik}\right) J - X_b J = \widetilde{R} - J^* X_- A^2/a, \tag{2.29}$$

where

$$\widetilde{R} = \widetilde{g}\widetilde{E} - X_a A - X_b B f, \tag{2.30}$$

$$X_- = -\langle\widetilde{g}'\widetilde{E}\widetilde{E}^*\rangle. \tag{2.31}$$

When changing from (2.28) to (2.29), formula (2.21) was used. Formula (2.31) defines a new quantity $X_-$. It will be shown below that $X_- = X_a - X_g$ (see formula (2.45)). Let us examine Eq. (2.29).

The noise arising from the item $R$ in the right side of (2.29) changes many times on a scale of the order of $1/|g|$ in view of complexity of the field structure, therefore, the last term in (2.29) having "irregular" phase does not influence propagation of noises of this kind. This term may be essential for those large-scale field components, which are weakly varied by diffraction at the distances of the order of the nonlinear length. Such components can be associated, for example, with illumination inhomogeneity, if a field with a complex structure was obtained as a result of a plane wave diffraction by a statistically homogeneous transparency. Denoting the small-scale part of the noise by $\tilde{J}$ and the large-scale part by $\check{J}$, we write down equations for each of the noise types:

$$(\partial/\partial z + \Delta_{\perp}/(2ik) - X_b)\,\tilde{J} = \tilde{R}, \tag{2.32}$$

$$(\partial/\partial z + \Delta_{\perp}/(2ik) - X_b)\,\check{J} = -(A^2/a)\,X_{-}\check{J}. \tag{2.33}$$

It should be emphasized that it is a relative weakness of noise that permits consideration of equations in the linear approximation in noise, and only so, these equations turned out to be independent.

Let us examine the small-scale part of the noise $\tilde{J}$. Taking into account the second of Eqs. (2.5), we put down (2.32) in the form of

$$(\partial/\partial z + \Delta_{\perp}/(2ik))\,\tilde{J}/B = \tilde{R}/B. \tag{2.34}$$

Let us suppose that the contributions of the right side of Eq. (2.34) into the noise intensity $\tilde{J}$ are quasiindependent in the points separated by an interval greater than the diffraction length $L_d$. Then the total intensity of the noise can be presented in the form of of a sum of quasiindependent items

$$\tilde{\eta} = \langle |J/B\,(L)|^2\rangle = \sum_{z=L_d,\ 2L_d,\ 3L_d,\dots,\ KL_d} \langle |\tilde{R}\,(z)\,L_d/B\,(z)\,|^2\rangle, \tag{2.35}$$

where the nonlinear medium length is $L = KL_d$, $K \gg 1$. Considering that variations of $a$ and $b$ on a diffraction length are small, we replace summation by integration in (2.35)

$$\tilde{\eta} = \int_0^L dz L_d \, \langle \tilde{R}\tilde{R}^* \rangle / b \, (z). \qquad (2.36)$$

This formula gives a qualitative estimate of the intensity of the small-scale distortions of the approximate solution $\tilde{E}$.

In the particular case of a third-order nonlinearity $g(I) = i\gamma I$, the magnitude of $\langle \tilde{R}\tilde{R}^* \rangle$ is easily calculated. Integration over Re $f$ and Im $f$ gives

$$\langle \tilde{R}\tilde{R}^* \rangle / b = |\gamma|^2 (a^2 + 6ab + 2b^2). \qquad (2.37)$$

Note that for the case of complete field modulation ($a = 0$), analogous formula was obtained in [32]. For nonlinearity of saturation type ($g(I) = g_0/(1 + I/I_s)$, $I_s$ = const, $g_0$ = const), dependence of $\langle \tilde{R}\tilde{R}^* \rangle$ on the intensities $a$ and $b$ is discussed in [39].

Now we consider Eq. (2.33) for the long-wave noise $\check{J}$. First of all, we show that $X_- = X_a - X_b$ (see formulae (2.9), (2.10), (2.21), (2.31)). It should be noted that the long-wave noise $\check{J}$ behaves as a part of the plane wave with respect to the rapidly changing field $f$. The idea of the proof consists in utilization of the results of the previous sections of this chapter for the field of the form of

$$E = \mathcal{A} + Bf, \qquad (2.38)$$

where

$$\mathcal{A} = A + \check{J}, \quad |\check{J}| \ll |A|. \qquad (2.39)$$

Repeating the derivations from Sec. 2.1 for this case, we write down

$$\left(\frac{\partial}{\partial z}+\frac{\Delta_\perp}{2ik}\right)\mathscr{A}=X_aA+X_a\breve{J}+aX_a'\breve{J}+A^2X_a'\breve{J}, \tag{2.40}$$

where $X_a'(a,b)$ is the derivative of the function $X_a(a,b)$ with respect to the first argument. On the other hand, expanding the expression $g(I)E$ in Eq.(1.12) and then averaging, we get

$$\left(\frac{\partial}{\partial z}+\frac{\Delta_\perp}{2ik}\right)\mathscr{A}=X_aA+X_b\breve{J}-X_-\breve{J}^*\frac{A^2}{a}+O(\breve{J}^2). \tag{2.41}$$

Since formulae (2.40) and (2.41) were written for the field $\breve{J}$ of a general form, coefficients multiplying $\breve{J}$ and $\breve{J}^*$ must be equal to each other

$$X_b=aX_a'+X_a,\qquad X_-A^2/a=-A^2X_a'. \tag{2.42}$$

The relation to be proved follows from formulae (2.42), i.e.

$$X_-=X_a-X_b. \tag{2.43}$$

Taking this formula into account, we shall regard the function $X_-(a,b)$ below as the difference between the effective increments of the plane wave and the modulated component. We note that formula (2.42) obtained as a by-product expresses one increment through the other one and its derivative.

Let us use relation (2.43) and show that Re $X_a$ > Re $X_b$, i.e. Re $X_-$ > 0. Basing on formula (2.31), we shall consider Re $\bar{g}'$ < 0. It corresponds to amplification satutation. Using the definition of the averaging operation < > and denoting integration variables as $t$ = Re$\xi$, $s$ = Im$\xi$, we present (2.31) in the form of

$$X_-=-aX_a'=-\frac{a}{\pi}\iint\limits_{-\infty}^{\infty}dt\,ds\,(\exp(-t^2-s^2))\times$$
$$\times\,g'\left((\sqrt{a}+\sqrt{b}\,t)^2+bs^2\right)\cdot\left((\sqrt{a}+\sqrt{b}\,t)^2+bs^2\right).$$

Since positivity of Re $X_-$ for $b \ll a$ follows from the results of Sec. 1.4, we shall consider $b > 0$. Changing the integration variable $t \to u = t - \sqrt{a/b}$, we obtain

$$X_- = \pi^{-1} \iint_{-\infty}^{\infty} \exp\left(-(u - \sqrt{a/b})^2 - s^2\right) g'\left((u^2 + s^2) b\right) \cdot (u^2 + s^2)\, b\, ds\, du.$$

Changing from Cartesian to polar coordinates, in accordance with formulae $u = r\cos\alpha$, $s = r\sin\alpha$, we have

$$X_- = -\int_0^{\infty} r\, dr\, g'(br^2)\, br\, \mathcal{G}\left(2\sqrt{a/b}\, r\right) \exp\left(-r^2 - a/b\right), \qquad (2.44)$$

where

$$\mathcal{G}(\beta) = \frac{1}{2\pi} \int_{-\pi}^{\pi} d\alpha\, (\exp(\beta\cos\alpha))\, (1 - 2\sin\alpha).$$

Using formula (3.915.4) from [91] at $\nu = 0$, $\nu = 1$, we get

$$\mathcal{G}(\beta) = I_0(\beta) - (2/\beta)\, I_1(\beta),$$

where $I_0$ and $I_1$ are the modified Bessel functions. Using representations (8.447.1), (8.447.2) (from [91]) of these functions in the form of power series, we obtain

$$\mathcal{G}(\beta) = \sum_{k=0}^{\infty} \left(\frac{\beta}{2}\right)^{2k} \Big/ k! - \frac{2}{\beta} \sum_{k=0}^{\infty} \frac{(\beta/2)^{2k+1}}{k!\,(k+1)!} \geqslant 0,$$

From the condition Re $g' < 0$ and (2.44), we get $\mathrm{Re}(X_a - X_b) = \mathrm{Re}\, X_- \geq 0$. It means that the plane wave is amplified with a greater increment than the modulated field component. Thus, the conclusion of Sec. 1.4 concerning the decrease of the field modulation depth in an amplifying medium should be spreaded to the case of a finite modulation depth. The difference between the effective increments tends to zero as $a \to 0$

(weak plane wave). In this case the medium seizes to distinguish between a plane wave and other spatial Fourier components of a field, and all the field components are amplified with the same efficiency.

Note that the positivity of Re $X_-$ was established for nonlinearity of a general type. Only the condition of the local increment decrease with intensity was used.

Let us return to the estimation of the long-wave noise intensity. The investigation of solutions of the linearized equation (2.33) will be carried out in analogy with Sec. 1.4. We expand the long-wave noise $\check{J}$ in real harmonics, i.e. present it in the form of $\check{J} = \sum_p J_p \cos(px + \varphi_p)$. The component with the transverse wavenumber $p$ satisfies the equation

$$\left(\frac{d}{dz} + \frac{ip^2}{2k} - X_b\right) J_p = -X_- J_p^* A^2/a. \quad (2.45)$$

This equation is a complete analogue of Eq. (1.38) written for the case of a weak modulation.

In the case of a purely amplifying medium, Re $X_- > 0$, and so large-scale perturbations of a plane wave grow slower than the modulated component of the basic field. In a saturable absorber ($g < 0, g' > 0$, $X_- > 0$), the long-wave amplitude perturbations are absorbed less efficiently than the modulated component of the basic field. The relative intensity of the long-wave noise can grow. In the case of low $p$, we obtain an estimate for a relative intensity of such noise, neglecting diffraction

$$\check{\eta} \approx \langle \check{J}\check{J}^* \rangle / b = \frac{\langle |\check{J}(0)|^2 \rangle}{b(0)} \exp\left(-2\int_0^L \mathrm{Re}\, X_- \, dz\right). \quad (2.46)$$

In the limiting case $b \ll a$, formulae (2.8), (2.43), and (2.21) give $X_- = -a g'(a)$, so estimate (2.46) agrees with formula (1.46) obtained in the frameworks of the perturbation theory. Taking into account Eqs. (2.9), (2.10) and formula (2.43), we can rewrite estimate (2.46) in the form of

$$\hat{\eta}(L) = \hat{\eta}(0)\,\mu(L)/\mu(0), \tag{2.47}$$

where $\mu(z) = b(z)/a(z)$.

Note that the growth of the relative intensity of such noise for a nonlinear absorption cannot be as essential as for a self-focusing instability because of limitations on the region of efficient interaction of the components associated with variations of their intensities.

In the presence of a self-focusing nonlinearity, the phase of the right side of Eq. (2.45) varies, and efficient amplification occurs for the components with the same rate of phase variations. Therefore, we try the solution of this equation in the form of $J_p = A\mathcal{B}\sqrt{b/a}$. Taking into account (2.7), we obtain

$$d\mathcal{B}/dz = -i\,(q + \operatorname{Im} X_-)\,\mathcal{B} - X_-\mathcal{B}^*, \tag{2.48}$$

where $q = p^2/(2k)$, the same as in Sec. 1.4. For $X_- = -i\gamma' a - \bar{g}' a$, this equation coinsides with Eq. (1.30) obtained for the case of weak modulation in the limit $b \to 0$. In particular, for a purely refractive nonlinearity, when Re $X_- = 0$, Im $X_- < 0$, Eq. (2.48) has an exponentially growing solution with an increment

$$\varkappa_q = ((\operatorname{Im} X_-)^2 - (q + \operatorname{Im} X_-)^2)^{1/2}. \tag{2.49}$$

In the case of joint influence of a self-focusing nonlinearity and amplification, intensity of the large-scale noise can be estimated from integration of the increment $\varkappa_q$ along the amplifier length

$$\eta_q(L) = \eta_q(0)\exp\left(2\int_0^L \varkappa_q\,dz\right).$$

Since the spatial spectrum of the large-scale perturbations is usually unknown, we put down the estimate of the large-scale perturbation relative intensity assuming that the noise in each cross-section grows with maximum (over all values of transverse wavevector) increment $\varkappa_{max} = -\mathrm{Im}\,X_-$:

$$\hat{\eta}(L) \approx \hat{\eta}(0)\exp(2\Phi(L)), \tag{2.50}$$

where

$$\Phi = \Phi(z) = \arg(B/A), \qquad d\Phi/dz = \mathrm{Im}\,(X_b - X_a) = -\mathrm{Im}\,X_-. \tag{2.51}$$

Analogy between this estimate and estimate (1.43) should be noted. The definition of dephasing $\Phi$ from formula (2.51) corresponds to the definition from formula (1.43) suitable for the case of a third-order nonlinearity in the limit $b/a \rightarrow 0$. Formula (2.50) leads to over-estimation, because it does not take into consideration the variations of the transverse wavenumber corresponding to the least stable mode. These variations were discussed in Sec. 1.4.

Estimates (2.36), (2.46), and (2.51) show that for small diffraction length of the field and in the absence of long-wave Fourier components, the noise is small. In this case, formulae (2.1), (2.8) quantitatively describe transformation of the spatial field structure in a medium with a weak nonlinearity. The most strict limitations on the validity of this description are associated with a plane wave instability for a self-focusing nonlinearity, in view of exponential growth of the noise intensity (formula (2.50)).

Besides, in this section, the condition Re $\chi_a$ > Re $\chi_b$ is proven to hold for the function $g$ of a general form at Re $g' < 0$, i.e. the plane wave is amplified more efficiently than the completely modulated component of the field.

## 2.4. Transformation of a partially modulated field in a homogeneous medium with a weak nonlinearity of a general form (conclusions of Ch.2)

Paraxial beam in a nonlinear medium is described by Eq.(1.12). If boundary effects are neglected, the approximate solution of Eq.(12) in the form of linear combination of a plane wave and a completely modulated component satisfying the linear diffraction equation corresponds to a partially modulated beam with a complex structure. For a weak nonlinearity, the primary beam distortion reduces to the variation of relation between the coefficients $A(z)$ and $B(z)$ in expression (2.1). This distortion is determined by the system (2.8) of two ordinary differential equations. Coefficients $\chi_a$ and $\chi_b$ in this system have the sence of effective increments of the plane wave and the modulated field component. They depend only on average intensities $a$ and $b$ of these components (see (2.9), (2.10)), therefore, system (2.11) is also closed. This system can be obtained also from the assumption that the field structure has the form of (2.1) and from the variation laws of the form of (1.26) not using the diffraction equation (1.12) explicitly.

The variations of the modulation depth $\mu = |B/A|$ and the dephasing $\phi = \arg(B/A)$ occurs in complete analogy with the case discussed in Sec.1.4 in the approximation $b \ll a$, system (2.8) determines the image contrast transformation (Fig.2.1) in the case of a finite modulation depth.

The difference between the approximate solution and the exact one corresponds to a certain field $J$ (noise). This field is a combination of a large-scale noise $\check{J}$ determined by the initial field components with the diffraction length greater or of the order of the nonlinear length $1/|g|$ and of a small-scale noise $\tilde{J}$, appearing in the nonlinear medium. The relative intensity $\check{\eta}$ = $<\check{J}\check{J}^*>/b$ of the small-scale noise is proportional to the field diffraction length and to the modulus squared of the resudue arising from substitution of the approximate solution (2.1) into Eq.(1.12). Formula (2.36) expresses this intensity through the solution $A(z)$, $B(z)$ of system (2.8). The relative intensity $\check{\eta}$ = $<\check{J}\check{J}^*>/b$ of the large-scale noise is approximately expressed by formula (2.50). The growth of this noise corresponds to the plane wave instability. At large dephasing $\Phi$ values, this noise (and particularly its components with the diffraction length of the order of the nonlinear length) limits the validity of the approximate solution (2.1).

If noise is neglected, the nonlinear distortions of the partially modulated wave with a complicated structure are regular. They are determined by system (2.8). These distortions may be eliminated with the help of amplitude and phase filtration using Zernike filter as shown in Fig.2.1. In the next chapter, solutions of system (2.8) for particular cases are discussed and compared with numerical solutions of Eq.(1.12).

# Chapter 3

## Interaction of a spatially modulated wave with a plane wave: examples, numerical estimates and numerical experiment

In this chapter, solutions of equations of coupled waves derived in Ch.1 are investigated on examples of certain intensity dependencies of the local increment. Numerical estimations are made in application to real media. Approximate solutions of the two-dimensional diffraction equation constructed in the frameworks of the speckle-mode approximation are compared with numerical solutions of this equation.

Formulae of the previous chapter were written for a medium with local nonlinearity of a general form. They answer the question of radiation structure transformation in the form of a system of two nonlinear differential equations, the right sides of these equations being expressed by double integrals (over real and imaginary parts of the quantity $f$). This representation is quite good for practical computations (these computations are much simpler than direct solution of the diffraction equation); however, from observation of the equations without computations, it is rather difficult in many cases to get the idea of the actual form of solution. In this connection, it is desirable to illustrate these equations by certain examples for the simplest cases. Such cases are nonlinearity of saturation type and third-order nonlinearity. Concretization of local increment $g(I)$ dependence enables representation of effective increments $X_a$ and $X_\ell$ using level lines and construction of the graphs of the solutions of the coupled wave equations (Sec. 3.1). Dependence of dephasing on nonlinearity parameters is examined in Sec. 3.2. Concretization of not only nonlinearity type but also actual field realizations permits verification of the results of the speckle-mode approximation using direct numerical computations (Sec. 3.3).

## 3.1. Decrease of the field modulation depth in amplifying medium

Let a beam with a random spatial modulation propagate in a homogeneous amplifier. Amplification saturation leads to a decrease of the local increment in the regions with high intensity values (in the presence of nonlinear refraction, the increment is complex, then its real part is implied). So the field modulation depth decreases.

This effect was discissed in [90] for the case, when the amplifier length $L$ is considerably smaller than the diffraction length $L_d$. However, for a field with complex structure, inverse relation usually holds, $L_d \ll L$. Let the nonlinear length $L_n$, at which the nonlinearity noticeably influences the field structure, also exceeds $L_d$. Decrease of modulation depth of the field with complex structure in an extended amplifier is described by system (2.8). Let $g(I) = g_0/(1 + I/I_s)$. For simplicity, we suppose $g_0 > 0$ (although the formulae remain valid for complex $g_0$ as well). Then expressions (2.9) and (2.10) take the form of

$$X_a(a,\ b) = (g_0 I_S/(2a)) \int_0^\infty e^{-r}\,(1 - (1 - (br - a)/I_S)/\sigma)\,dr,$$
$$X_b(a,\ b) = (g_0 I_S/(2a)) \int_0^\infty e^{-r}\,(1 - (1 + (br - a)/I_S)/\sigma)\,dr, \qquad (3.1)$$

where

$$\sigma = (1 + 2\,(a + br)/I_s + (a - br)^2/I_s^2)^{1/2}.$$

This case is considered in detail in [39], and so we do not dwell on it in the present paper.

### 3.2. The case of purely refractive nonlinearity

To elucidate the influence of the refractive nonlinearity upon the features of an image carried by radiation, let us examine the case, when amplification is switched off. In this section, solutions of system (2.8) for purely refractive media are discussed.

In a purely refractive medium, $\mathrm{Re}\, g = 0$, and the effective increments $X_a, X_b$ also take purely imaginary values. In this conditions, $a$ = const, $b$ = const, the coefficients $A$ and $B$ are varying only in phase, and the solution of system (2.8) has the form of $A = A(0)\exp(X_a z)$, $B = B(0)\exp(X_b z)$. The dephasing $\Phi = \arg(B/A)$ is proportional to the nonlinear medium length

$$\Phi = \mathrm{Im}\,(X_b - X_a)\,z = -\mathrm{Im}\,(X_-)\,z. \qquad (3.2)$$

The same can be said about the relative intensity of the small-scale noise

$$\tilde{\eta} = \langle \tilde{J}\tilde{J}^* \rangle / b \approx (L_J \langle \tilde{R}\tilde{R}^* \rangle / b)\, z. \qquad (3.3)$$

This fact simplifies the analysis essentially.

In Sec. 1.4, dependence of the contrast of an image carried by the field on the dephasing $\Phi$ was discussed. It was pointed out that the strong dependence of the contrast on the dephasing in the vicinity of $\Phi = \pi/2$ enables one to utilize qualitative features of the image for quantitative measurement of the nonlinear medium parameters. Let us consider this possibility.

We suppose that a plane wave with intensity $\bar{I}$ illuminates an amplitude transparency as shown in Fig. 2.1. The field just behind the transparency may be written in the form of $A(0) + B(0) f_0(x,y)$, where $A(0) = A_0 \bar{I}$, $B(0) = B_0 \bar{I}$, $f_0(x,y)$ is a real function of the transverse coordinates, and $\langle f_0 \rangle = 0$, $\langle f_0^2 \rangle = 1$. Positive coefficients $A_0$ and $B_0$ are determined by the transparency.

Let in the case of a linear medium (in the limit of small intensities) the field

$$E(x, y, L) = A(L) + B(L)\, f(x, y, L) \qquad (3.4)$$

at the nedium output can be obtained by setting the field in the transparency plane:

$$E_0(x, y) = A(L) + B(L) f_0(x, y). \quad (3.5)$$

A wide-aperture lens produces on a screen the field distribution (with uncertain regular phase, which is quadratic in distance from the optical axis)

$$E_0(-Mx, -My)/M, \quad (3.6)$$

where $M = L_2/L_1$ is the magnification coefficient for the transparency imaging on the screen. For a linear medium, $A$ = const, $B$ = const, therefore, at the point of the screen corresponding to the transparency coordinates $x$, $y$, the intensity is

$$I_{xy} = | A(L) + B(L) f_0(x, y) |^2/M^2. \quad (3.7)$$

Thus, we found the connection between the intensity distribution on the screen and the field distribution at the output of the nonlinear medium, using the linear case.

Returning to the nonlinear case, we rewrite (3.7) in the form of

$$I_{xy} = (a(L) + 2(a(L) b(L))^{1/2} \cos(\Phi(L)) f_0(x, y) + b(L)(f_0(x, y))^2)/M^2. \quad (3.8)$$

It should be remembered that contrary to the dephasing $\phi$, the function $f_0(x,y)$ here depends neither on the nonlinear medium length nor on the intensity, it is completely determined by the transparency. It follows from (3.8) that at $\phi = 0$, the image on the screen is positive, and the points corresponding to the transparency regions with high transmission are illuminated better. At $|\phi| = \pi$, the image is negative (the transparency regions with high transmission look dark). In Sec. 1.4, negatives and positives in a certain plane $z$ = const were discussed as well, but the lens was not supposed to be placed between the medium and the image. The form of (3.7) reveals the identity of these two cases. In fig.2.1, the screen and the transparency may be interchanged (it is only necessary to take measures that the lens intersepts the greater part of the beam behind the transparency). This interchange does not lead to modifications of the following formulae, reasonings and conclusions (the model example in Fig.1.3 is also applicable to both cases). Therefore, the contrast reversal observed in [92] can also be treated as a manifestation of dephasing between the field components (in the conditions of [92] the image of a regular net appeared just in the nonlinear medium due to the Talbot effect).

At $|\Phi| \approx \pi/2$, small variation of intensity or medium properties (for example, nonlinear refraction constant) results in quantitative variation of the image features. Note that even at small $\mu$, the image contrast may be noticeable. For instance, at $\mu = 0.04$, the contrast is $\mathcal{V} = 0.2$. Since even a naked eye is able to observe relative brightness variations $\Delta I / I \approx 0.01$, this effect may be used for measurements of nonlinear parameters.

Let us consider the case of purely refractive saturable non-linearity in a greater detail:

$$g(I) = i\gamma I_S \left(1 - \frac{1}{1 + I/I_S}\right) = g_0 - g_0 \frac{I_S}{(I + I_S)}, \quad g_0 = i\gamma I_S \qquad (3.9)$$

This dependence corresponds to a dye with separation between the transition frequency and the field frequency somewhat greater than the Lorentz broadening of the spectral line. The ratio of real and imaginary parts of the increment is proportional to this detuning (see Sec.1.4 and Refs.[77-79]). The linear refractive index $n|_{I=0}$ is chosen as the constant part of the refractive index for convenience (see Sec.1.1), so that $g(0) = 0$. Let the magnitudes of $\gamma$ and $I_S$ be

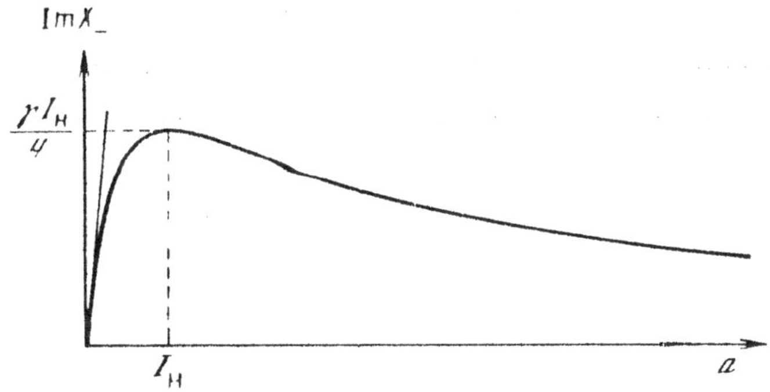


Fig. 3.1. Dependence of Im $X_- = d\Phi/dz$ on the plane wave intensity in a medium with saturable refractive nonlinearity of the form of (3.9) for small modulation depth. The dependence was constructed in accordance with formula (3.10).

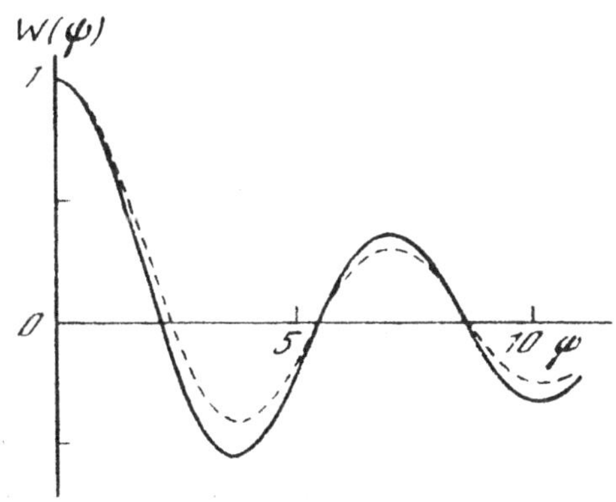


Fig. 3.2. Dependence of $W$ value (see formula (3.13)) characterizing averaged image on dephasing extremum value for two cases: Gaussian envelope $h(t) = \exp(-\pi t^2/\tau^2)$ (solid curve) and hyperbolic envelope $h(t) = 1/\mathrm{ch}(\pi t/\tau)$ (dashed curve).

the ones to be determined. Varying the nonlinear medium length $L$ and measuring the intensity, at which the image of an amplitude transparency becomes the phase one, it is possible to find out the dependence of Im $X_- = \gamma/(2L)$ on $a = A_0^2 I$. Restricting ourselves to small modulation depths, we expand formulae (3.1) in $\mu$. Here we have

$$\operatorname{Im} X_- = \gamma_a/(1 + a/I_s) + O(b). \tag{3.10}$$

This dependence is shown in Fig.3.1. The position of the maximum on the curve gives the magnitude of $I_s$ and its height gives the magnitude of $\gamma$. The constant $\gamma$ may also be determined from the curve slope near the origin (for example, if the initial intensity is not sufficient to trace the whole curve).

Intensities inducing a noticeable nonlinearity of a medium are usually achieved in short pulses. This fact hinders observation of image contrast variations with intensity varying. Fast temporal scanning or special shutter transmitting only the most intense part of an optical pulse are needed. So it is interesting to consider the situation, when intensity in every point of the screen (see Fig.2.1) is averaged over the passage time of an optical pulse. In this case the quantity $W_{xy} = \int I_{xy}\,dt$ is observed (or otherwise recorded), where $I_{xy} = |A + Bf_0(x,y)|^2$.

Let us show that the qualitative features of even an averaged image may be used for measurements of $\gamma$. For simplicity, we restrict ourselves to the case of a third-order nonlinearity ($a \ll I_s$). Let us present the intensity of the averaged image in the form of

$$W_{xy} = \left(A_0^2 W_0 + 2A_0 B_0 W f_0(x, y)\right) \bar{I}/M, \tag{3.11}$$

neglectong the term of the order of $b$ in (3.8). Here

$$W_0 = \int \overline{I}(t)\, dt/\check{I}, \quad W = \int \overline{I}(t)\cos(\Phi)\, dt/\check{I}, \quad \check{I} = \max \overline{I}(t). \tag{3.12}$$

Direct observation enables determination of at least the sign of $W$, if not its numerical value. $W > 0$ and $W < 0$ correspond to a positive and a negative, respectively. $W = 0$ corresponds to the lack of an image (in the last case the term of the order of $b$ that was neglected may result in a pale contouring).

In a medium with a third-order nonlinearity, the dephasing is $\Phi = \gamma a L$. Let the optical pulse be bell-shaped: $\overline{I}(t) = \check{I} h(t)$, where $0 < h(t) < 1$. The magnitude of $\check{I}$ can easily be controlled with linear optical filters preserving the function $h(t)$. In this case

$$W = W(\psi) = \int h(t)\cos(\psi h(t))\, dt, \tag{3.13}$$

where $\psi = \check{I} A_0^2 L$ is the extremum value of the dephasing. For smooth envelopes $h(t)$, the function $W(\psi)$ is oscillating. Pulses with exotic envelopes having a sharp maximum and decreasing slowly at the infinity do not give zeroes of the function $W$ and are inconvenient for the measurements utilizing the proposed method. Let us number the zeroes of the function $W$. Every $\psi_m$ corresponds to an intensity $\check{I}_m$ value, at which the transverse modulation of an averaged image vanishes. From the values of, for instance, $\psi_1$, $\check{I}_1$, the constant

$$n_2 = \gamma n_0/k, \quad \gamma = \psi_1/(L A_0^2 I_1). \tag{3.14}$$

may by calculated.

For Gaussian and hyperbolic envelopes

$$h(t) = \exp(-\pi t^2/\tau^2), \quad h(t) = 1/\mathrm{ch}(\pi t/\tau), \quad \tau = \mathrm{const}, \tag{3.15}$$

the function $W(\Psi)$ is presented in Fig.3.2. The curves are close to each other, and the method is insensitive to the envelope shape. It is only important that it is smooth and rapidly decreasing as $t \to \pm\infty$. If the envelope is unknown, the values of $\Psi_n/(LA_0^2)$ can be measured using a medium with a known constant $n_2$. In [72] carbon disulphide was used as a reference medium, but in the experimental setup, interferometer and temporal scanning device were essential. Neither of them is needed in the method proposed above. The method used in [73,77,78] resembles the proposed method most of all techniques described in literature. In these works, efficiency of a plane wave self-diffraction by a light-induced grating was measured to calculate the constant $n_2$.

Let us consider an example showing the actual possibilities of image contrast transformation in a self-focusing medium and utilization of this phenomenon for measuring $n_2$. Let $k \approx 10^5$ cm$^{-1}$, $L \approx 10$ cm, $n_0 \approx 1$, $n_2 \approx 10^{-14}$ cm$^2$/W, $\gamma \approx 10^{-9}$ cm/W. It is seen from (3.13) that in this case, the dephasing $\Phi = \gamma a L \approx \pi/2$ is provided at intensities $a = n_0\Phi/(kn_2L) \approx 10^8$ W/cm$^2$. In these conditions (see (2.50)), intensity of initial large-scale components increases by a factor of $e^{2\Phi} \approx 23$. If the size of the small-scale field inhomogeneities is $\ell_\perp \approx 0.001$ cm ($L_d = k\ell_\perp^2 \approx 0.1$ cm), the beam width $L_\perp \approx 10\,\ell_\perp L/L_d \approx 1$ cm is sufficient to neglect the edge effects, i.e. power of the order of $10^8$ W suffices to perform the measurements. The relative intensity of the small-scale noise (see (2.36), (2.37)) is $\ell_\perp \approx (kn_2/n_0)^2 L_d L a^2 \approx 0.01$, so it should not increase the uncertainty of the method. This uncertainty is determined by the lowest contrast permitting an eye (or other sensor) to distinguish a positive image from a negative one. The relative uncertainty is probably about several percent. If the light source power is insufficient, the measurements can be made at $|\Phi| < \pi/2$ as well. In this case, a phase mask should be placed in the Fourier plane (see Fig. 2.1), and the uncertainty of the result increases. For defocusing medium ($\gamma < 0$), the instability limiting the possibilities of observation of large $\Phi$ is absent, so the uncertainty of the proposed method can be decreased, if the measurements are performed at $\Phi = -3\pi/2$ ( or even at $\Phi = -5\pi/2$). This method may be helpful in investigation of variations of properties of amplifying media for small (within the linewidth) frequency shifts of the light being amplified.

Note that for illustration of transitions between positive and negative described above and also for their utilization in information processing, the required beam power may be decreased approximately by a factor of billion due to utilization of artificial nonlinear media based on liquid crystals. In particular, in [93] conditions were studied, in which effective nonlinear permittivity was $\mathcal{E}_2 \approx 0.1$ cm /erg, i.e. $\gamma \approx kn_2 \approx h\mathcal{E}_2/c \approx 1$ cm/W (instead of $10^{-9}$ cm/W in the estimation presented above).

Thus, dephasing of the field components in the presence of refractive nonlinearity of the medium leads to more essential distortions of a beam structure than variations of the modulation depth do. The reversal associated with this dephasing may be used for investigation of the features of nonlinear medium response. In the next section, model examples of the contrast reversal in a refractive medium are examined.

## 3.3. Numerical experiment

The basic equations and formulae of the previous chapter ((2.1), (2.8)-(2.11), (2.30), (2.36)) are based on the statistical hypothesis saying that the averagings of the form of $\langle \Phi(f) \rangle_\perp$ over transverse coordinates are close to the corresponding average magnitudes over the ensemble of field $f$ realizations. These averages are calculated in the assumption that the field distribution in every point of space is a complex Gaussian one. Examination of particular cases in this chapter is also based on the statistical hypothesis. It is interesting to compare the conclusions of the previous section with the results obtained using some other method.

In this section, approximate solution (2.1) of Eq. (1.12) is compared with numerical solution for several one-dimensional ($E(x,y,0)$ does not depend on $y$) random realizations of initial field. Although in this case (when the field depends on only one transverse coordinate) the phenomenon of self-focusing collapse is absent (see Sec. 1.3), the effects associated with variations of intensity and modulation depth and also with the plane wave instability are quite analogous to the actual three-dimensional case.

Let us change to dimensionless coordinates through a formal substitution $k = 1$. The reverse transition to dimensional quantities is performed using formulae

$$\begin{aligned} x_{\text{dim}} &= Mx/k, & z_{\text{dim}} &= M^2 z/k, & g_{\text{dim}} &= kg/M^2, \\ (l_\perp)_{\text{dim}} &= Ml_\perp/k, & L_{\text{dim}} &= ML/k, & E_{\text{dim}} &= kE/M^2, \end{aligned} \tag{3.16}$$

where $M$ is a scaling coefficient. If the correlation radius $l_\perp$ of the field $E$ is of the order of unity, Eq. (1.12) in dimensionless coordinates

$$(\partial/\partial z - (i/2)\,\partial^2/\partial x^2)\,E = g\,(EE^*)\,E \tag{3.17}$$

is adequate for $M \gg 1$ (at $M \lesssim 1$ the condition of paraxiality is violated). In particular, the calculations described below may correspond to real fields and media for $100 \lesssim M \lesssim 1000$.

Eq. (3.17) was solved using expansion in spatial Fourier components (like in [20], but there the field was two-dimensional and had a relatively simple structure). Performing Fourier transformation according to the formula

$$E = E(x, z) = \frac{1}{\sqrt{2\pi}} \int \exp(ipx - ip^2z/2)) E_p(z)\, dp, \tag{3.18}$$

we get

$$\frac{\partial E_p}{\partial z} = \frac{1}{\sqrt{2\pi}} \int \exp(ipx - ip^2z/2)\, R(x, z)\, dp, \tag{3.19}$$

where

$$R(x, z) = R = g(EE^*)\, E. \tag{3.20}$$

For numerical analysis of $E(X,Z)$, $E(Z)$ were presented on a uniform net of $N$ nodes with an increment $\Delta X = \Delta p = \sqrt{2\pi/N}$. The net size $L_\perp = N\Delta X$ determined the field periodicity along $X$ axis. In the frameworks of the parabolic approximation, at $g = 0$ the field $K$ would be periodic along $Z$ axis as well, and from equation $(\Delta pN/2)^2 T = \pi$ the corresponding Talbot period $T = 2N$ could be found. The calculations described below are carried out for $Z < T$, so there was no influence of periodicity. One of criteria of validity of the chosen net is the condition

$$|E_p|^2 \ll \Delta x \int_{-1/\Delta x}^{1/\Delta x} dq\, |E_q|^2 \quad \text{when} \quad |p| > 1/\Delta x. \tag{3.21}$$

This is the condition of smallness of the net increment in comparison with the field inhomogeneity size in the momentum representation. On the chosen net, Eq.(3.19) transforms into a system of $N$ ordinary differential equations of the form of

$$dE_p/dz = R_p(E(z)). \tag{3.22}$$

Integrals in formulae (3.21), (3.22) were replaced by finite sums of a descrete Fourier transform. These sums were computed using the fast Fourier transform algorythm [94, p. 402]. System (3.22) was resolved using Runge-Kutt scheme of the fourth order (see [95, p. 692]:

$$E_p(z+\Delta z) = E_p(z) + \frac{\Delta z}{6}\left(R_p^1 + 2R_p^2 + 2R_p^3 + R_p^4\right) + O\left(\Delta z^5\right), \qquad (3.23)$$

where

$$R_p^1 = R_p(E(z)), \qquad R_p^2 = R_p\left(E(z) + \Delta z \cdot R^1/2\right),$$
$$R_p^3 = R_p\left(E(z) + \Delta z \cdot R^2/2\right), \qquad R_p^4 = R_p\left(E(z) + \Delta z \cdot R^3\right). \qquad (3.24)$$

In the computation process, the magnitudes of

$$A = \langle E\rangle_\perp, \qquad B = \langle Ef^*\rangle_\perp, \qquad J = E - A - Bf \qquad (3.25)$$

were computed for comparison with the results of the statistical approach. To test the adequacy of representation of Eq. (3.17) on a finite net, conservation (or variation, if the medium was not purely refractive) of integrals of the form of (1.17) valid for a one-dimensional periodic field was examined in the process of calculation. These are: average intensity

$$\overline{I} = \frac{1}{L_\perp} \int_{-L_\perp/2}^{L_\perp/2} EE^*\, dx, \qquad (3.26)$$

average momentum

$$\overline{P} = \frac{1}{L_\perp} \int_{-L_\perp/2}^{L_\perp/2} \mathrm{Im}\left(E\frac{\partial E^*}{\partial x}\right) dx = \frac{1}{L_\perp}\int pE_pE_p^*\, dp, \qquad (3.27)$$

average Hamiltonian

$$\overline{H} = \frac{1}{L_\perp} \int_{-L_\perp/2}^{L_\perp/2} (|\partial E/\partial x|^2 + U)\, dx, \tag{3.28}$$

where

$$U = \int_0^{EE^*} \operatorname{Im} g(I)\, dI. \tag{3.29}$$

In a purely refractive medium, the calculated values of $\overline{I}$, $\overline{P}$, $\overline{H}$ were conserved with high accuracy. Their relative variations did not exceed 1% even when a self-focusing instability developed.

Boundary conditions in a cross section $z$ = const were either introduced by hand or composed with a random number generator. In the latter case, the distributions of $E_p$ components were taken independent at $\rho_0 < |p| < \rho$ ; $E_p = 0$ at $|p| > \rho$ and at $0 \geq |p| < \rho_0$; $E_0 = \sqrt{N}A(0)$. The values of $E_p$ at $p \neq 0$ were normalized so that $\sum_p E_p E_p^*$ $= N\, b(0)$. In the calculations described below, the parameters were $\rho_0 \approx 1$, $\rho \approx 5$; $a(0)$ and $b(0)$ were varied. The function $f$ in (3.25) was determined from the formula

$$f(x, z) = \frac{1}{\sqrt{2\pi}} \int_{p\neq 0} \exp(ipx - ip^2 z/2)\, E_p(z)\, dp. \tag{3.30}$$

Features of the chosen spatial Fourier component spectrum of the initial function $f$ determined the initial form of the selective correlation function

$$\begin{aligned} c(x, z) &= \frac{2\pi}{L_\perp} \int_{-L_\perp/2}^{L_\perp/2} (E(y, x) - A(z))(E(y+x, z) - A(z))^*\, dy = \\ &= 2\pi L_\perp^{-1} \int_{p\neq 0} e^{ipx} E_p(z)\, E_p^*(z)\, dp \end{aligned} \tag{3.31}$$

in the plane $z = 0$. For this plane, a nonstrict replacement of $E_p(0) \cdot E_p^*(0)$ by its average value

$$\langle E_p E_p^* \rangle = \begin{cases} \dfrac{b(0)\, N\, \Delta p}{2(s - s_0)}, & \text{if } s_0 < |p| < s \\ 0 & \text{in opposite case} \end{cases} \quad (3.32)$$

reduces (3.31) to.

$$c(x,\ 0) \approx b(0)\, N\, \Delta p/(2(s - s_0)\, L) \int_{s_0 \leqslant |p| < s} e^{ipx}\, dp \approx b(0)\, \frac{\sin(sx)}{sx}. \quad (3.33)$$

The ratio $s_0/s \approx 0.2$ being small, the correlation functions for such realizations turned out to be like the functions in the right side of (3.33) (integral in (3.31) was replaced by sum). The transverse modulation scale was determined from equation $C(\ell_\perp, 0) = C(0,0)/2$. It has a solution $\ell_\perp \approx 1.9/s \approx 0.95$. From this, the diffraction length is $L_d = \ell_\perp^2 \approx 3.6/s \approx 0.144$. This value is used below for comparison of the noise intensity $\langle JJ^* \rangle$ with the estimation from formula (2.33).

To test the algorythm of numerical solution of system (3.18)-(3.21), we consider an example of calculation of propagation of a localized beam with a simple structure (Fig. 3.3) in a medium with nonlinearity of the form of

$$g(I) = g_0/(1 + I) + i\gamma I. \quad (3.34)$$

The initial field in the plane $z = 0$ was real, the intensity profile had a shape of a somewhat asymmetric bell (curve 1 of Fig. 3.3,d; a small part of the net is depicted). For the calculation with a coarse step $\Delta z = 0.2$ at $g_0 = 1$, $\gamma = 1$, the approximations of $\tilde{I} = L_\perp I$, $\tilde{H} = L_\perp H$, $\tilde{P} = L_\perp P$ (see Sec. 1.3) are presented by circles. The solid curves correspond to estimations from formulae of the form of (1.18) (the boundary conditions in the plane $z = 0$ are taken from formulae (1.17)).

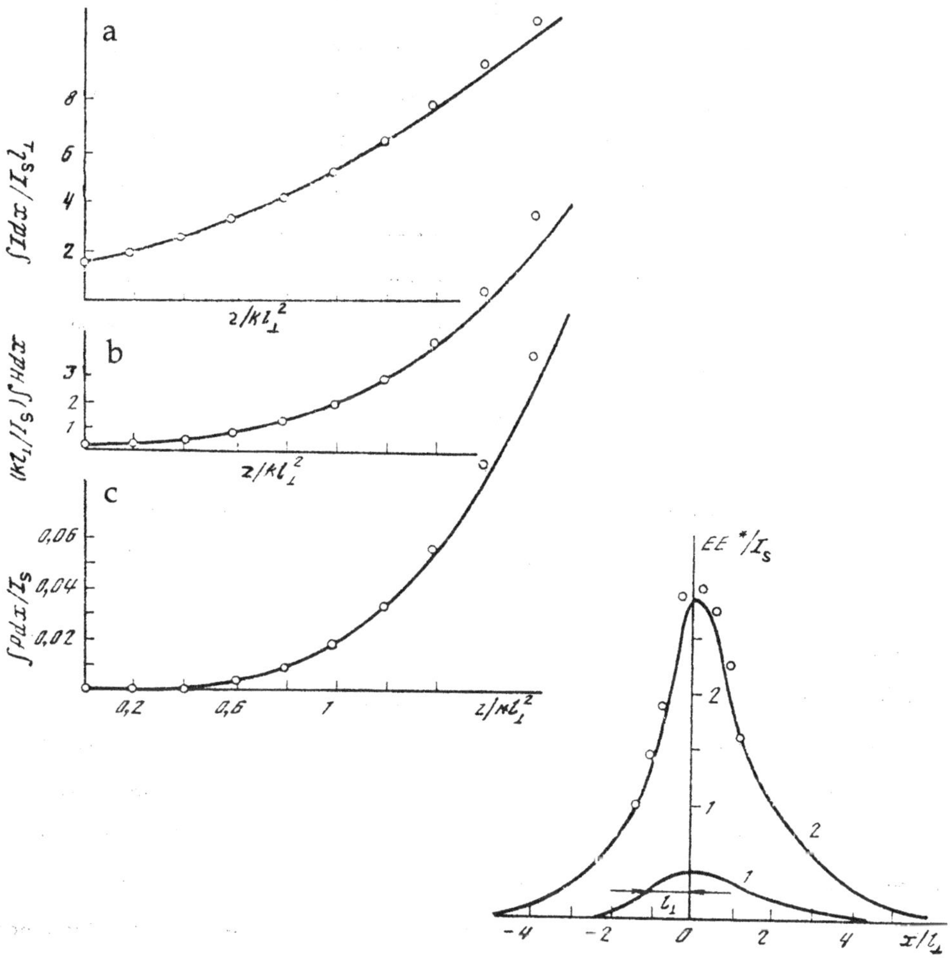

Fig. 3.3. Variation of the Noether integrals (a–c) of one-dimensional field, which is real in the cross section $\bar{z} = 0$ in an amplifying self-focusing medium; distributions of intensity (d) of this field in the cross sections $\bar{z} = 0$ (1), $\bar{z} = 1.6$ (2).

Solid curves were obtained by integration of formulae of the form of (1.18), circles correspond to definitions of the Noether integrals (1.17). This calculation was performed on a net consisting of 64 nodes, curve 2 in fig.3.3,d is a solution obtained on a finer net.

The magnitudes of these integrals calculated for solutions with a smaller step ($z \leq 0.1$, the calculations were also performed on finer nets) practically coinside with the solid curves, although the solution with a coarse step differs from the exact one already in the plane $z = 1.6$. It indicates that in the calculations of the beam parameter transformations in a nonlinear medium, integration of expressions of the form of (1.18) gives essentially more exact results than the direct computation of the necessary integrals over approximate field distribution at the nonlinear medium output.

In the calculation described above, the field has a simple structure, it is impossible to select the plane wave and the randomly modulated component. Let us proceed to the description of the results of the numerical experiment with fields having a complex structure.

In a purely amplifying medium ($g_o > 0$), dependence of the intensities $a$ and $b$ on the longitudinal coordinate was well described by system (2.11) at $0 < g_o L_d < 1$. Variation of the intensities $a$ and $b$ and the noise relative intensity $\langle JJ^* \rangle / b$ in the process of amplification is presented in Fig.3.4 for $g_o = 5$, $a(0) = 0.03$, and $b(0) = 0.07$. Dashed curves corresponding to system (2.11) and estimation (2.36) are given in the same graphs for comparison.

It can be noted that the intensities $a$ and $b$ obtained from the numerical calculation are somewhat delayed with respect to the theoretical (from system (2.11)) values, since system (2.11) does not account for the amplification saturation by the noise field. Insignificant variation of the correlation function corresponds to smoothening of the initial rectangular distribution of the spatial Fourier component intensity. The noise energy spectrum does not coinside exactly with the initial field spectrum.

Analogous calculations were performed for saturable absorber at $g_o < 0$ (solution of the form of (2.1) for this case was discussed in

[96] using Fourier component description). The correspondence between numerical and approximate results was essentially worse, on the whole. Already at $g_0 = -2$ ($g_0 L_d = -0.6$), variations of the modulation depth turned out to be irregular and lagged behind the theoretical curve. The noise intensity determined from (2.22) became underestimated. Typical dependencies of the intensities $a$ and $b$, the modulation depth $\mu = b/a$, and the noise intensity on the longitudinal coordinate are presented in Fig. 3.5. The instability is probably associated with the field modulation depth increase, when the large-scale amplitude perturbations of the plane wave (arising from the small-scale field in nonlinear interaction) are amplified in comparison with the modulated component. The assumption about the smallness of the field diffraction length is not valid in these conditions. Note that the resonant absorption instability does not lead to the complete destruction of the field structure, since nonlinear interaction of the field components stops, when the average field intensity becomes small.

In Figs. 3.4 and 3.5, results are given only for the intensities $a$ and $b$. A phase shift acquired by the coefficients $A$ and $B$ is very small in these calculations.

In a self-focusing medium ($\mathrm{Im}\, g_0 < 0$), the plane wave is unstable. One may hope that system (2.8) is adequate on the ground that the initial field (0) does not contain Fourier components, which could be efficiently amplified at $\gamma a \approx p^2/2$. Until these components are not formed due to nonlinear interaction and amplified, the primary nonlinear distortion of the field is the dephasing between the plane wave and the completely modulated field component. For $\mathcal{S}_0 \approx 1$, $\mathcal{S} \approx 5$, and $a \approx 1$, the noise becomes significant at $\phi \approx 4$. In Fig. 3.6a, b, one of realizations of initial (in the plane $z = 0$) distribution of the field amplitude and phase is presented. Here $a(0) = 0.5$, $b(0) =$

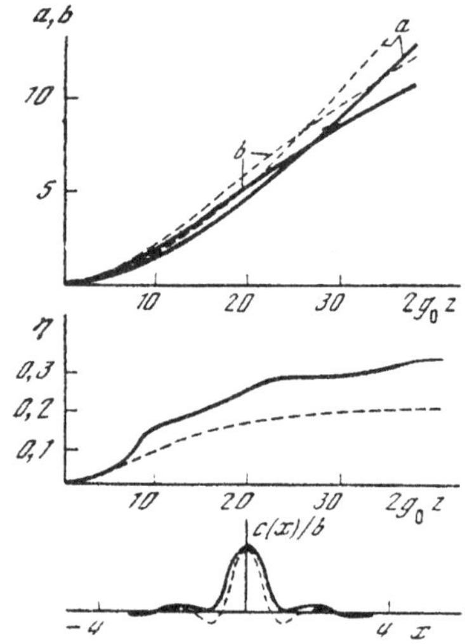


Fig. 3.5. Dependence of the intensities $a$, $b$ the modulation depth $\mu = b/a$ and the noise relative intensity $\eta$ on the saturable absorber length $-g_0 z$ at $g_0 = -2$, $a(0) = b(0) = 5$.

Solid curves are the result of the numerical experiment, dashed curves correspond to solution og system (2.11) and estimation (2.36). In the upper graph, the dashed curves and the corresponding solid curve coinside.

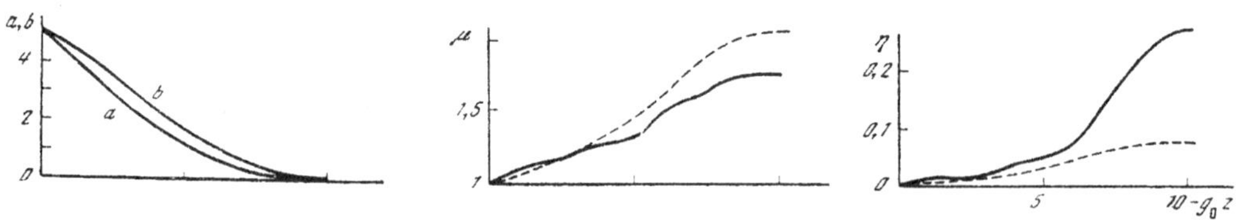


Fig. 3.4. Dependence of intensities $a$, $b$ and the noise relative intensity $\eta$ on the amplifier length $2g_0 z$ at $g_0 = 5$, $a(0) = 0.03$, $b(0) = 0.07$ (solid curves are the result of the numerical experiment, dashed curves correspond to solution of system (2.11) and estimation (2.36)); correlation function (formula (3.31)) in the cross sections $z = 0$ (dashed curve) and $g_0 z = 15$ (solid curve).

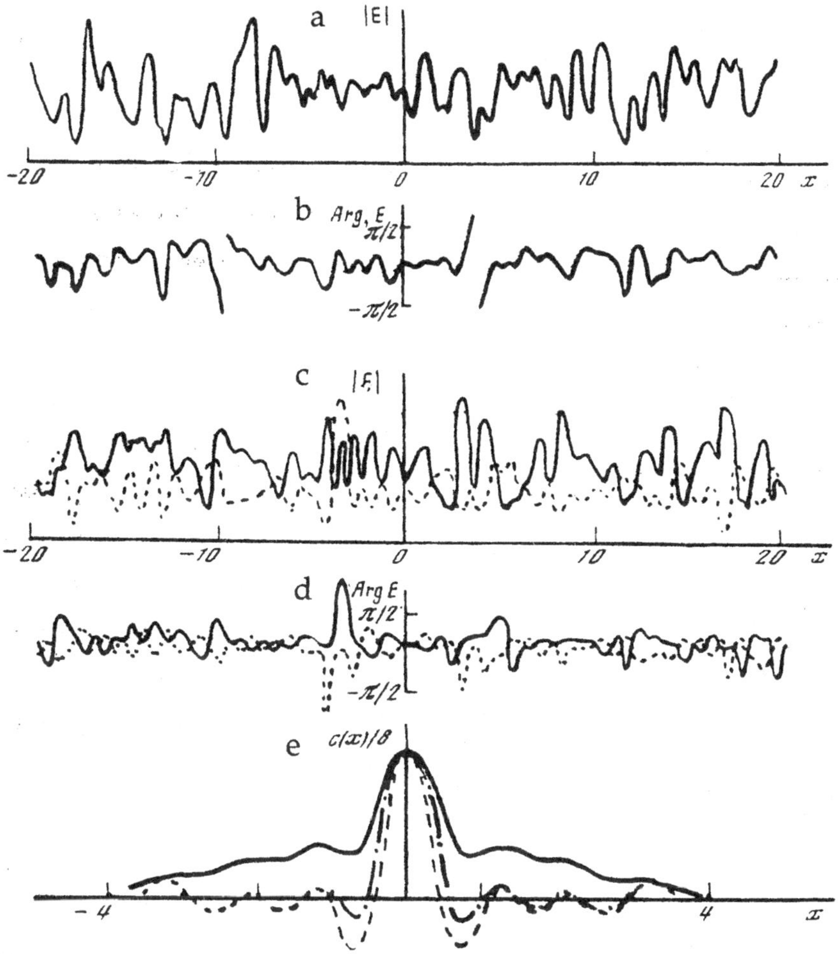


Fig. 3.6. Dependence of the amplitude and the phase (a-d) in one of the field realizations on the transverse coordinate in the cross sections $z$ = 0 (a, b) and $z$ = 22 (c, d) in comparison with the distributions, which would have been in the same cross sections at $g$ = 0 (dashed curves); the correlation function (e) in the cross sections $z$ = 0 (dashed curve), $z$ = 22 (dash-dotted curve), and $z$ = 32 (solid curve).

0.2. Dependence of the dephasing and the noise intensity for the realization at $g_0$ = -1 is depicted in Fig.3.7. Until $\Phi$ < 2, the dephasing and the noise intensity grow almost linearly. But at $\Phi \approx 2$, instability exhibits, and the noise starts to grow faster borrowing intensity from the plane wave. Its decay delays the dephasing. In all the performed calculations, the condition $\eta \ll 1$ turned out to be violated at $\Phi$ = 4.

System (2.8) gives a solution $A(z) = \sqrt{0.5}\exp(-0.564iz)$, $B(z) = \sqrt{0.2}\exp(-0.426iz)$ for this case. Estimation of the small-scale noise from formulae (2.36), (2.37) gives $\tilde{\eta} \approx 0.004z$. The long-wave component generation efficiency was not estimated analytically, but one can conclude from Fig.3.7 that in this case $\check{\eta} \approx 0.005ze^{0.27z}$. The exponent magnitude agrees with formula (2.46). In realizations with lower values of the parameter $S_0$ restricting the long-wave field component (but with the same $a(0)$, $b(0)$), the instability manifested itself at lower $\Phi$ values.

Consideration of actual realizations shows that the field carries a negative image at $|\Phi| = \pi$, in accordance with the statement of Sec.1.4. In Fig.3.6c,d, the field amplitude distribution (of the same realization as in Fig.3.7) in the cross section $z$ = 22 is also presented (for $z$ = 22 the dephasing is equal to $\pi$) in comparison with the distribution that would be in the same plane at $g$ = 0. In contrast to Fig.1.3, where the contrast reversal of the real field is presented, here the field is complex, and so the phase distribution is also given. The average value of Arg($E$) is close to zero, and this result is fortuitous to some degree. The matter is that for $a$ = 0.5, $b$ = 0.5, the dephasing $\Phi = \pi$ is achieved almost in the same plane, where the increment of the argument of $A$ is equal to $4\pi$.

It can be noted that the field phase looks as if it has changed the sign, i.e. Arg $E|_{g=0}$ has a minimum where Arg $E|_{g_0=-i}$ had a maximum in

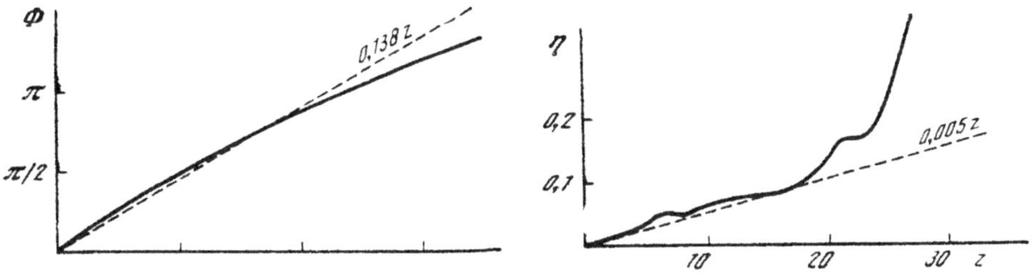


Fig. 3.7. Dependence of the dephasing $\phi$ and the noise relative intensity $\eta$ on the self-focusing medium length $z$ at $g_o = -i$, $a(0) = 0.5$, $b(0) - 0.2$.

Dashed straight lines correspond to solution of system (2.11) and estimation (2.36).

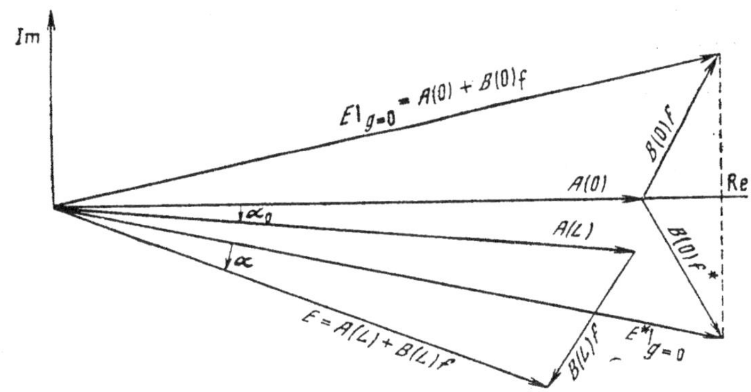


Fig. 3.8. Diagram illustrating phase reversal in contrast reversal of a field with a small modulation depth in a medium with nonlinear refraction.

Vector $E$ corresponds to field in a certain point of the cross section $z = \pi / \mathrm{Im}(X_a - X_b)$, vector $E|_{g=0}$ corresponds to the field that would be in the same point in the absence of nonlinearity. The phase difference of the vectors $E = A(L) + B(L)f$ and $E^*|_{g=0} = A(0) + B(0)f^*$ is expressed by the angle $\alpha = \mathrm{Arg}(E) - \mathrm{Arg}(E^*|_{g=0})$. For a small modulation depth (i.e. at $A << |Bf|$), $\alpha \approx \alpha_0 = \mathrm{Arg}(A(L)/A(0))$. The angle $\alpha$ is almost independent of $f$, and so is practically constant in the beam cross section. Therefore, the field $E$ phase behaves almost the same way as the field $E^{**}|_{g=0}$ phase, i.e. $\mathrm{Arg}(E) \approx -\mathrm{Arg}(E|_{g=0}) + \mathrm{const}$. In this sense, the field phase changed the sign.

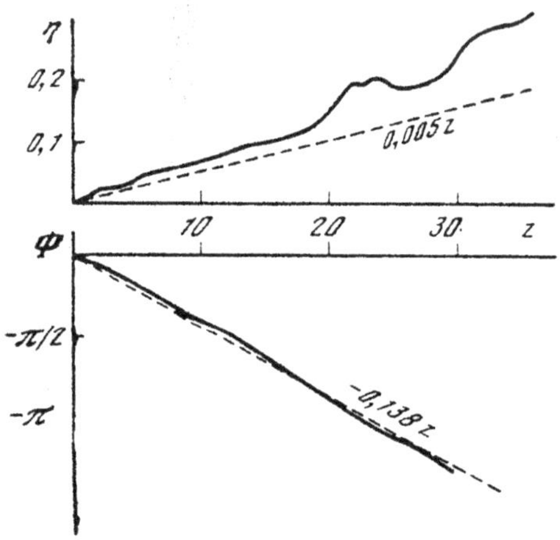


Fig. 3.9. Dependence of the noise intensity $\eta$ and the dephasing $\Phi$ on the self-focusing medium length at $g_o = -i$.
Dashed straight lines correspond to solution of system (2.11) and estimation (2.36). The boundary condition in the cross section $z = 0$ is the same as for the cases in Figs. 3.6, 3.7.

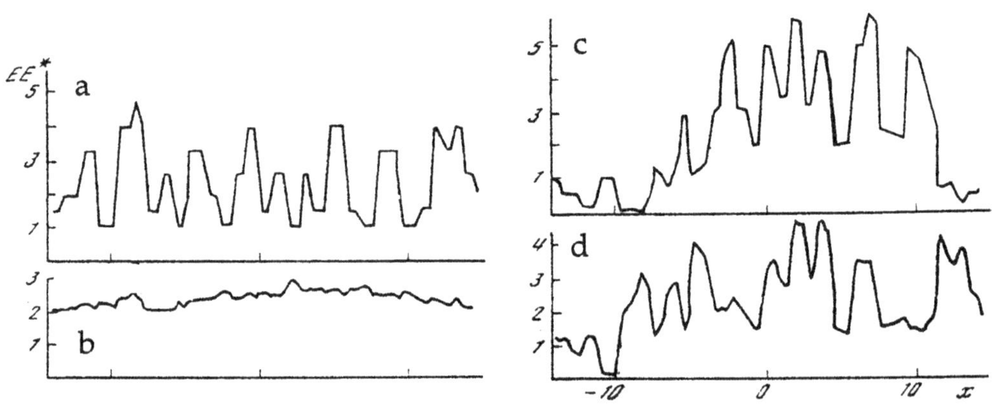


Fig. 3.10. Distribution of the field intensity in the cross section $z$ = -10 corresponding to the screen (1) plane in Fig. 2.1 at $\gamma a(0)L = 0$ (a), $\pi/2$ (b), $\pi$ (c), $3\pi/2$ (d).
Here $a(0) = 2.26$, $b(0) = 0.12$, $\gamma = 0.02$. Dependence of the phases of the coefficients A, B of this field on the length $L$ is presented in the upper part of Fig. 3.11.

the linear case (and vice versa). The cause of this phenomenon can be understood from the phase diagram (Fig.3.8). For small modulation depth, the phase difference between the field $E|_{g_0=-i}$ and the field $E|_{g=0}$ is practically constant across the beam.

Analogous calculation was carried out for $g_0 = i$, it corresponds to a defocusing medium (fig.3.9). In contrast to the previous case, the plane wave is stable, and so the dephasing as well as the noise intensity varies almost linearly at large $-\phi$ $(-\phi \gtrsim \pi)$ as well. At $= -2\pi$, we get a positive image again, i.e. the field in the plane $z = 44$ is almost the same as it would be in this plane at $g = 0$. Instability was not exhibited also for the complete field modulation, when $\phi = 0$.

To illustrate applicability of the speckle-mode approximation to the fields having initially non-Gaussian statistics, analogous calculation was performed for the field, which was real in the plane $z = -10$ and propagated in a medium with a third-order nonlinearity, $g(I) = i\gamma I$, $\gamma = 0.02$. This calculation was carried out on a net consisting of $N = 128$ points. The field amplitude was introduced by hand without any special order. The only requirement was the absence of large jumps from point to point, which resulted in a slow decrease of Fourier components $E_p$ at large $|p|$. Intensity (amplitude squared) of this field is presented in Fig.3.10,a. This figure corresponds to the plane $z = -10$ or the transparency plane, if we consider that the experiment is modelled in accordance with the diagram shown in Fig.2.1. For this field, $a(0) = 2.26$, $b(0) = 0.12$.

Dependence of phases of the coefficients $A$, $B$ on the medium length is presented in the upper part of Fig.3.11 for $\gamma = 0.02$. The phases are growing almost linearly until $\Phi < 2$. At $L = 35$ the image becomes a contour one, and intensity in the image plane is almost independent of the transverse coordinate (Fig. 3.10, b). At $L = 70$, the image on the screen is negative (Fig. 3.10, c), though it is somewhat distorted, in contrast to the ideal case in Fig. 1.4. At greater $L$, the image does not already correspond to the initial one either in amplitude or in phase (Fig. 3.10, d), the "pattern" is destroyed and its reconstruction is possible only through an inverse nonlinear transformation, for example, integrating system (3.25) with a negative step (this test calculation has been performed).

Results of analogous calculation for a defocusing nonlinearity ($\gamma = -0.02$) are shown in the lower part of Fig. 3.11 and in Figs. 3.12, a–e. Here, as well as in Fig. 3.9, the linear phase evolution is not perturbed, the contrast reversal is close to the ideal one, the noise intensity can be qualitatively described by formula (2.36), which for this case (see formula (2.37)) gives $\tilde{\eta} \approx 0.0004z$ (for this realization, $\ell_{\perp} = 0.4$, $L_d = 0.16$).

With reserves concerning instability, the results obtained in the previous sections are supported by direct numerical calculations. Simplicity of the statement under test gave possibility to avoid averaging of the numerical experiment results over a great number of field realizations, which was performed, for instance, in [21]. For amplifying and for defocusing media, the uncertainty of the approximate (obtained from system (2.8) and based on the statistical hypothesis about field randomness) solution (2.1) of the nonlinear diffraction equation did not exceed several percent, in agreement with estimate (2.36). In a self-focusing medium with weak nonlinearity, the structure of the modulated component can be preserved at $\Phi < \pi$. For greater $\Phi$, appearance and rapid growth of the long-wave spatial Fourier components makes system (2.8) inadequate. Note that measure-

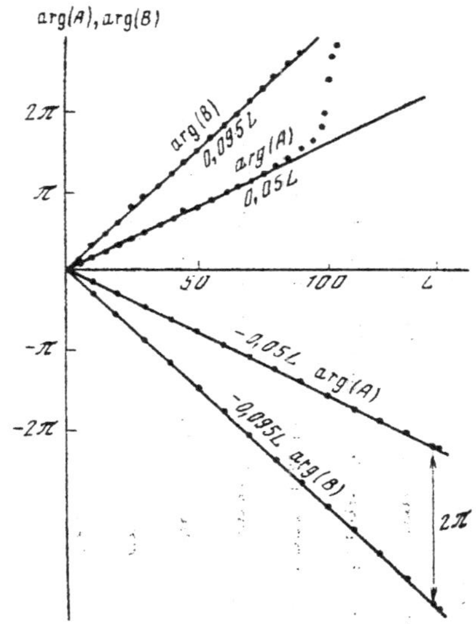


Fig. 3.11. Dependence of the phases of the coefficients A, B on the length $L$ of the medium with a third-order nonlinearity at = 0.02 (the upper part of the graph) and at $\gamma$ = -0.02 (the lower part of the graph). In these calculations, $a$ = 2.26.

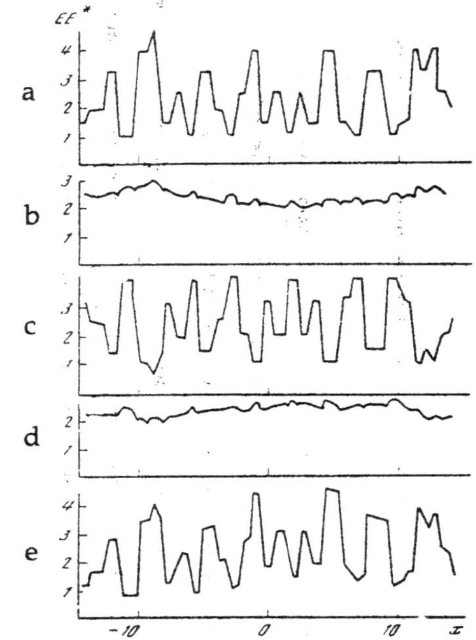


Fig. 3.12. Distribution of the field intensity in the cross section $z$ = -10, corresponding to the screen (1) plane in Fig. 2.1 at $\gamma a(0)L$ = 0 (a), $-\pi/2$ (b), $-\pi$ (c), $-3\pi/2$ (d), $2\pi$ (e).
Here $a(0)$ = 2.26, $b(0)$ = 0.12, $\gamma$ = -0.02. Dependence of the phases of the coefficients A, B of this field on the length $L$ is presented in the lower part of Fig. 3.11.

ments of the constant with the technique proposed in Sec. 2.4, dephasing $\Phi = \pi/2$ is sufficient.

## 3.4. Image contrast transformation and validity of the statistical hypothesis (conclusions of Ch. 3)

Concretization of nonlinearity type permitted definite numerical estimations to be made concerning actual physical media. Examination of actual field realizations supports general conclusions made on the basis of the speckle-mode approximation in the frameworks of hypothesis about statistical (random) field distribution along the transverse coordinates.

Interaction of the modulated field component with the plane wave in an amplifier with saturation leads to a field modulation depth decrease (saturation in an absorbing medium leads to a growth of the modulation depth). The corresponding variation of the intensities of the plane wave and the modulated field component can be found from Figs. 3.2, 3.3; the relative intensity of the noise arising from the medium nonlinearity may be estimated from Fig. 3.4. These estimates do not contradict experimental data.

The presence of nonlinear refraction may lead to image contrast reversal. This effect provides a technique for measuring medium nonlinear parameters, in particular, the nonlinear refraction constant (formulae (3.4), (3.11) and Fig. 3.1). Numerical estimates indicate practical applicability of this technique.

Approximate solutions of the diffraction equation (1.12) of the form of (2.1) are compared with exact solutions for several realizations of the initial field. The best agreement of this approximate solution with the results of the numerical experiment is observed for amplifying and defocusing media. For absorbing media, instability is observed, which was discussed in Sec. 2.3. For saturable absorbers, this instability results in a somewhat greater relative noise intensity growth than estimation (2.36) predicts. Finiteness of an effective interaction region limits the development of this instability. In the case of a self-focusing medium, the instability develops until the image carried by the field is destroyed. In the performed calculations, this destruction was observed for dephasing values of $3 < \Phi < 4$. This values increase with ratio of diffraction length $L_d$ and nonlinear length $L_n$ decreasing and is particularly sensitive to the long-wave part of the modulated component of the initial field.

The calculations were carried out for fields with both Gaussian and non-Gaussian statistics in the initial cross section. At $L_d \ll L$, no quantitative difference between these two cases was observed.

The calculations and estimates performed show that in the case of a partial field modulation, the image contrast transformation in non-linear media can essentially influence the amplified field structure.

Note that the contrast reversal for unidirectional radiation propagation is associated with nonlinear medium refraction. In a purely active medium, the ratio $B/A$ does not change the sign. However, in the presence of colliding (mutually conjugated) plane waves, the contrast reversal may be observed for a real increment as well. This case is a subject of the following chapter.

# Chapter 4.

## Colliding beams with complicated structure in an amplifier; formation of negative in a purely amplifying medium

In this chapter, transformation of radiation structure in nonlinear interaction of colliding modulated light beams is investigated. Effect of negative image formation due to spatially nonuniform saturation of the amplifying medium is considered from a general point of view as well as on a model example (a field with a simple structure in a short amplifier).

As we pointed out in Sec.1.1, active systems are often used in practice with radiation passing through the amplifier in mutually opposite directions. Interaction of counter-propagating beams is described by system (1.8). The purpose of this chapter is the construction of approximate solutions of this system and their interpretation. The simplest is the case, when amplification is concentrated on a short interval, so that diffraction may be neglected. This case was analyzed in [90] for unidirectional propagation of radiation. Analogous consideration for interacting colliding beams is presented below in Sec.4.1 as a model example. In amplification of complex images, the opposite case (when amplification on a diffraction length is small) is more actual. In assumption that colliding beams with complicated structure efficiently overlap in an amplifier, mutually conjugated components may be selected from these beams and their interaction may be observed. For description of this interaction in holography [34-36] and phase conjugation [27-33], theory of coupled waves is used (this theory was used in two previous chapters in the simplified version). Effects of degenerate four-wave interaction are more or less essential in an amplifier, depending on its nonlinearity. For its description, equations of coupled waves are written in Sec. 4.2; in Secs.4.3 and 4.4 they are analyzed in application to a purely active nonlinearity (without nonlinear refraction). Metal vapor lasers with the frequency of the amplified field coinsiding with the amplification spectrum line are the example of such active media. Since in vapor-phase active media the amplification saturation cannot be considered strictly local (see Sec.1.1), it is important to estimate the sensitivity of the effects discussed in Secs.4.3, 4.4 to locality of the medium response. Sec.4.5 is devoted to this problem.

## 4.1. Amplifier with small length (example of formation of negative)

In this section, an example illustrating the process of a negative image formation in an amplifier with nonlinearity of saturation type is considered. Let amplification be concentrated on a short interval. Besides, instead of inequalities (2.7), relation $L_d \gg L$ holds. Let colliding waves with intensities $I_1$ and $I_2$ propagate in a medium with nonlinearity of saturation type. Neglecting diffraction, we have

$$dI_1/dz = 2F(I_1, I_2)\, I_1, \tag{4.1}$$
$$-dI_2/dz = 2F(I_2, I_1)\, I_2, \tag{4.2}$$

where the function $F$ is determined from formula (1.10). Boundary conditions are imposed at the opposite ends of the nonlinear medium:

$$I_1(0) = I_0(x, y) = I_0, \qquad I_2(L) = I_L(x, y) = I_L. \tag{4.3}$$

Coordinates $x$, $y$ enter formulae (4.1)-(4.3) only as parameters. Let it be $I_1 \gg I_2$ everywhere in the amplifier. Then

$$\begin{aligned} F(I_1, I_2) &= F(I_1, 0) = g(I_1) = g_0/(1 + I_1), \\ F(I_2, I_1) &= F(0, I_1) = -g'(I_1) = g_0/(1 + I_1)^2. \end{aligned} \tag{4.4}$$

Here, as well as in Sec. 3.3, intensity is measured in units that determine saturation (in (1.10), $I_s$ should be set equal to 1). With allowance for (4.4), we rewrite system (4.1), (4.2) as

$$dI_1/dz = 2g_0 I_1/(1 + I_1), \tag{4.5}$$
$$-dI_2/dz = 2g_0 I_2/(1 + I_1)^2. \tag{4.6}$$

Separating the variables, we get (see, for instance, [81])

$$I_2(0)/I_1 = I_1(L)(1 + I_0)/((1 + I_1(L))I_1). \qquad (4.7)$$

This formula illustrates the influence of the first beam transverse structure upon the second beam structure. Eq.(4.7) takes a particularly simple form, when $I_1(0) = I_0 \ll 1$, $I_1(L) \ll 1$, and the initial intensity of the second beam does not depend on the transverse coordinates, i.e. $I_2(L) = I_L$ = const. In this case

$$I_2(0) = I_L/I_0. \qquad (4.8)$$

The most intense regions of the transverse profile of the first wave correspond to the least intense regions of the image, and vice versa.

In Fig.4.1, dependence of a certain input field $I_1(0) = I_0(x,y)$ on the transverse coordinate $x$ is given. We can assume that a certain cross section $y$ = const is implied, or else $dI_0/dy = 0$. A weak counter-propagating wave without transverse modulation is supposed to enter the amplifier. Passing from the plane $z = L$ to the plane $z = 0$ it acquires a transverse structure. This structure is graphed in Fig. 4.1 according to formula (4.8). The figure shows that the counter-propagating wave acquires the structure negative with respect to the initial wave structure. The negative formation here is a manifestation of a spatially inhomogeneous burning of the medium inversed population

Formulae (4.1)-(4.8) and Fig.4.1 were obtained in the assumption that $L_d \gg L$. Propagation of a complex signal in an extended amplifier corresponds to another limiting case, $L_d \ll L$. In subsequent sections, a negative is shown to develop at $L_d \ll L$ as well.

### 4.2. Equations for coefficients, the case of colliding beams

Interaction of colliding paraxial beams in a medium with local

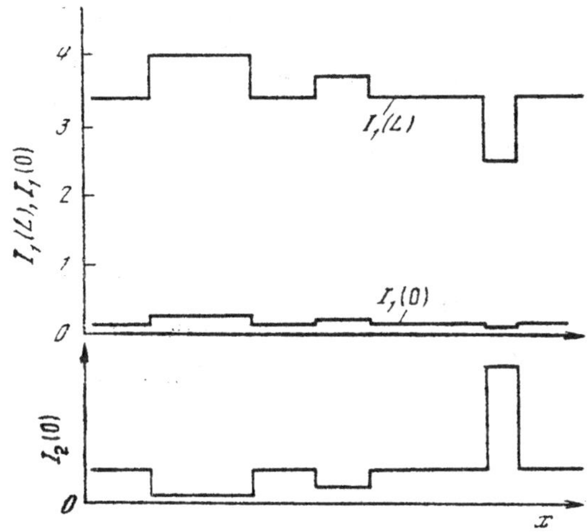


Fig. 4.1. Transverse intensity distribution of the basic signal $I_1$ at the input of a short amplifier $I_1(0)$ (arbitrarily chosen) and at the output $I_1(L)$ (calculated in accordance with formula (4.7)); intensity distribution of the counter-propagating field $I_2$ in the cross section $z = 0$ (calculated in accordance with formula (4.9)).

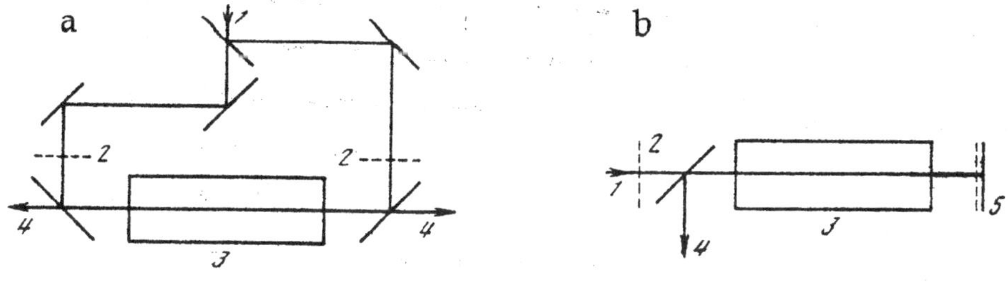


Fig. 4.2. Possible experimental layouts, in which colliding modulated beams propagate in the amplifier
(a) is the case in which boundary condition at $z = 0$ and $z = L$ may be determined independently; (b) is the case in which the field $E_1$ is related to the field $E_2$ at one end of the nonlinear medium.
1, initial beam; 2, transparencies with complicated structure; 3, nonlinear medium (amplifier); 4, output beam; 5, mirror (possibly, PC mirror).

response is described by system (1.8). In this section, approximate solution of the system is derived for wide beams with complicated spatial structure. In enables changing from partial differential equations to a system of ordinary differential equations, like it was made in Sec. 2.1 for a single beam.

To take into account a possible presence of mutually conjugated components in colliding beams, we utilize the speckle-mode theory of coupled waves developed in [28-33]. We try approximate solution of system (1.8) in the form of

$$E_1(x,\ y,\ z) = A_1 + B_1 f_1 + C_1 f_2^*, \qquad E_2(x,\ y,\ z) = A_2 + B_2 f_2 + C_2 f_1^*, \quad (4.9)$$

where

$$\left(\frac{\partial}{\partial z} + \frac{\Delta_\perp}{2ik}\right) f_1 = 0, \qquad \left(-\frac{\partial}{\partial z} + \frac{\Delta_\perp}{2ik}\right) f_2 = 0,$$

besides, $\langle f_1 f_1^* \rangle_\perp = \langle f_2 f_2^* \rangle_\perp = 1$, $\langle f_1 \rangle = \langle f_2 \rangle = \langle f_1 f_2 \rangle_\perp = 0$, and the coefficients $A_1, B_1, C_1, A_2, B_2, C_2$ depend only on $z$. The fields $f_1$ and $f_2$ are supposed to be random and independent. Such independence takes place, for instance, if the fields $E_1$ and $E_2$ were obtained by transmitting a plane wave through two different transparencies as shown in Fig. 4.2, a. In particular, boundary conditions $C_1(0) = 0$, $C_2(h) = 0$ correspond to the case of two completely independent transparencies. One transparency can suffice, if the beam after passing a nonlinear amplifier is reflected by a mirror. If the mirror is separated from the medium by an interval exceeding the diffraction length, the modulation of the colliding beams can be considered as if it were independent.

It is natural to choose coefficients in formulae (4.9) basing on the condition of minimum residue like in Sec 2.1 for $E_2 = 0$. Let us find the projections of the right sides of Eqs.(1.8) on every of three items in formulae (4.9). We get a system

$$
\begin{aligned}
dA_1/dz &= P_{a1} = \langle F(|E_1|^2, |E_2|^2)E_1\rangle, \\
dB_1/dz &= P_{b1} = \langle F(|E_1|^2, |E_2|^2)E_1 f_1^*\rangle, \\
dC_1/dz &= P_{c1} = \langle F(|E_1|^2, |E_2|^2)E_1 f_2\rangle, \\
dA_2/dz &= -P_{a2} = -\langle F(|E_2|^2, |E_1|^2)E_2\rangle, \\
dB_2/dz &= -P_{b2} = -\langle F(|E_2|^2, |E_1|^2)E_2 f_2^*\rangle, \\
dC_2/dz &= -P_{c2} = -\langle F(|E_2|^2, |E_1|^2)E_2 f_1\rangle
\end{aligned}
\tag{4.10}
$$

for the coefficients in expansion (4.9).

In the accepted denotations, the physical problem corresponds to imposing boundary conditions at the left end of the nonlinear medium ($z = 0$) for the coefficients with subscript 1 and at the right end of the nonlinear medium ($z = L$) for the coefficients with subscript 2. Like in Sec.2.2, we require that $A_1(0) \geq 0$, $B_1(0) \geq 0$, $A_2(L) \geq 0$, $B_2(L) \geq 0$. These limitations do not make the analysis less general. As for the coefficients $C_1$ and $C_2$, they cannot be given an arbitrary phase at the nonlinear medium border, since the freedom associated with the uncertainty of phases of the fields $f$ has been already utilized.

In assumption that averaging in the right sides of Eqs.(4.10) is performed over complex independent random quantities $f_1$ and $f_2$ distributed with the density of $\exp(-f_1 f_1^* - f_2 f_2^*)/\pi$, these equations form a closed system. It determines approximate solution (4.9) of system (1.8). After concretization of boundary conditions of a general form, system (4.10) may be solved, for example, using iterations. With allowance for the purpose of the present paper, it is interesting to analyze easily interpreted particular cases, when some of the coefficients in formulae (4.9), (4.10) are equal to zero. We note that system (2.8) describing transformtion of a single beam modulation depth is one of particular cases of system (4.10), when $A_2 = B_2 = C_1 = C_2 = 0$.

Let us examine the question of uncertainty of approximate solution (4.9) of system (1.8). In the absence of instabilities studied in [1-3,25,27], we may proceed in analogy with Sec. 2.2. Neglecting the possibility of amplification of large-scale perturbations, we present exact solution of system (1.8) in the form of $E_1 = \tilde{E}_1 + J_1$, $E_2 = \tilde{E}_2 + J_2$. The fields $J_1$ and $J_2$ are supposed to be independent. Then, considerations of Sec. 2.3 concerning the smallness of diffraction length in comparison with nonlinear length are extended to the present case and give the estimation for

$$\langle J_1 J_1^* \rangle = \int_0^L L_д \left( \exp \left( 2 \int_z^L \operatorname{Re} \tilde{X}(\xi) \, d\xi \right) \right) \langle \tilde{R}_1 \tilde{R}_1^* \rangle \, dz, \tag{4.11}$$

where

$$\begin{aligned} \tilde{R}_1 &= \tilde{F}_1 \tilde{E}_1 - \langle \tilde{F}_1 \tilde{E}_1 \rangle - \langle \tilde{F}_1 \tilde{E}_1 f_1 \rangle f_1 - \langle \tilde{F}_1 \tilde{E}_1 f_2 \rangle f_2^*, \\ \tilde{F}_1 &= F(\tilde{E}_1 \tilde{E}_1^*, \tilde{E}_2 \tilde{E}_2^*), \qquad \tilde{X} = \langle F'(\tilde{E}_1 \tilde{E}_1^*, \tilde{E}_1 \tilde{E}_1^*) \tilde{E}_1 \tilde{E}_1^* + F_1 \rangle. \end{aligned} \tag{4.12}$$

Here $F'(u,v) = dF(u,v)/du$ is a derivative of the function $F$ with respect to the first argument. Note that for $E_2 = 0$ (a single beam), $F_1 = F(E_1 E_1^*, 0) = g(E_1 E_1^*)$, and (3.7) reduces to (2.36).

Assumption of independence of the noises $J_1$ and $J_2$ and neglection of parametric instabilities in investigation of colliding wave amplification are partially justified by the fact that these instabilities are usually associated with self-focusing nonlinearity; they are described completely enough in the works on phase conjugation (PC). In purely amplifying media, PC does not lead to parametric instability because of low efficiency. Even creation of a laser with a self-pumped PC mirror proved to be a rather delicate experiment (see [84]), and in the works on image amplification, parasitic oscillations were observed only in the presence of additional scatterers.

So for random fields with complicated structures, approximate solution (4.9) of system (1.8) of partial differential equations (describing diffraction of paraxial colliding beams in a medium with local nonlinearity) is determined by system of six ordinary differential equations (4.10). Three particular cases are analyzed in two subsequent sections in application to calculation of amplifiers, namely, the completely modulated colliding waves in comparison with well-known (see, for example, [5]) case of two colliding plane waves (Sec. 3.3), and the completely colliding correlated beams, $B_2 = C_2 = 0$ (Sec. 4.4).

### 4.3. The case of two colliding plane waves in amplifier and the case of two colliding completely modulated waves

To comprehend the essence of phenomena occurring in amplification of colliding spatially modulated beams, it is helpful to analyze the case, when the modulation of the colliding waves is complete and independent, and to compare it with the case of colliding plane wave amplification. In this section, such analysis is carried out on an example of a saturation type nonlinearity, when the function $F(I_1, I_2)$ is determined from formula (1.10) ($I_s = 1$, like in Sec. 4.1).

If dependence of the fields $E_1$ and $E_2$ on the transverse coordinates is not essential, we assume $B_1 = C_1 = B_2 = C_2 = 0$ in formulae (4.9), (4.10). This is the case of two plane wave amplification considered in [5, 80]. For this case, system (4.10) gives

$$dA_1/dz = F(A_1A_1^*,\ A_2A_2^*)\,A_1, \qquad -dA_2/dz = F(A_2A_2^*,\ A_1A_1^*)\,A_2. \tag{4.13}$$

General solution of this system has the form of

$$1 + (P^2 + M^2)/2 + (1 + P^2 + M^2 + P^2M^2)^{1/2} = 2I_{\min} = \text{const},$$
$$\ln\frac{Pm - Mp}{Pm + Mp} - PM - 2g_0z = 2g_0z_0 = \text{const},$$

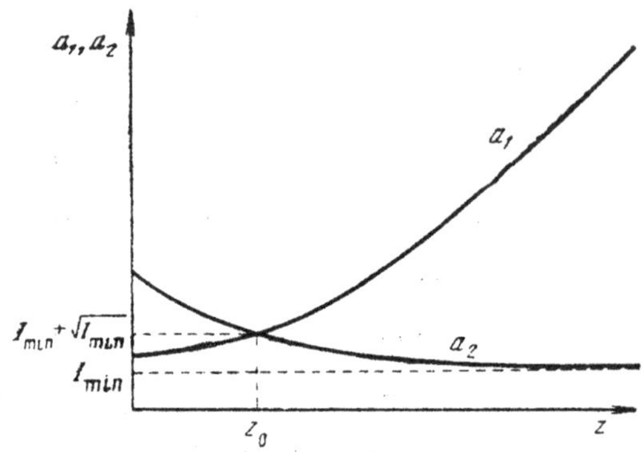


Fig. 4.3. Dependence of the intensities of colliding plane waves in the amplifier on the transverse coordinate.
As $z \to \pm\infty$, the intensity of one of the counter-propagating beams tends to the asymptote $I_{min}$.

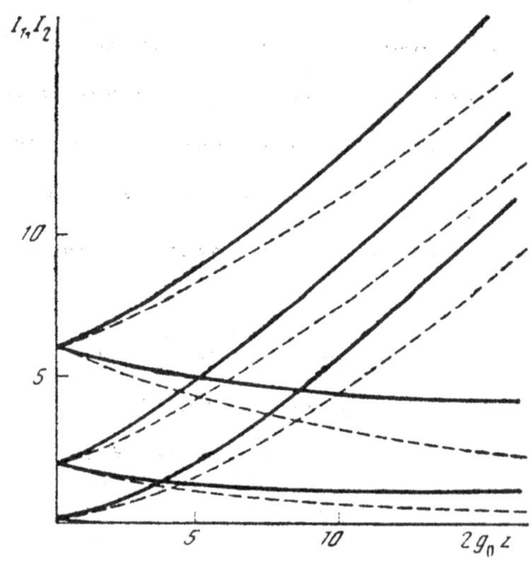


Fig. 4.4. Dependence of the intensities of the colliding waves on coordinate.
Solid curves correspond to the case of plane waves (solution of system (4.13)), dashed curves correspond to the case of modulated waves (solutions of system (4.16)).

where $P = A_1 + A_2$, $M = A_1 - A_2$, $m = 1 + (1 + M^2)^{1/2}$, $p = 1 + (1 + P^2)^{1/2}$. The meaning of the constants $I_{min}$ and $Z_0$ introduced here is clear from Fig. 4.3. $I_{min}$ is an asymptote of intensity at $Z \to \infty$; $Z_0$ is a cross section, in which intensities of the colliding waves are equal to each other.

Let us consider amplification of plane waves with complete and independent spatial modulation. At $A_1 = C_1 = A_2 = C_2 = 0$, we have

$$P_{b1} = (2\pi)^{-1} \int_{-\pi}^{\pi} \langle g(|B_1 f_1 + B_2 f_2 e^{i\theta}|^2) \cdot (B_1 f_1 + B_2 f_2 e^{i\theta}) f_1^* \rangle d\theta,$$

$$P_{b2} = (2\pi)^{-1} \int_{-\pi}^{\pi} \langle g(|B_2 f_2 + B_1 f_1 e^{i\theta}|^2) \cdot (B_2 f_2 + B_1 f_1 e^{i\theta}) f_2^* \rangle d\theta \quad (4.14)$$

(see (1.7), (4.10)). Performing change of variables analogous to (2.18) we get

$$P_{b1} = X_b(0, b) B_1, \qquad P_{b2} = X_b(0, b) B_2, \quad (4.15)$$

where $b = b_1 + b_2$, $b_1 = B_1 B_1^*$, $b_2 = B_2 B_2^*$ (see (2.10)). Substituting (4.15) into (4.10), we obtain a system for the average intensities $b_1$ and $b_2$

$$\frac{db_1}{dz} = 2\,\mathrm{Re}\,(0, b) \cdot b_1, \qquad \frac{db_2}{dz} = -2\,\mathrm{Re}\,X_b(0, b) \cdot b_2. \quad (4.16)$$

Its solution has the form of

$$b_1(z)\, b_2(z) = I_0^2 = \mathrm{const},$$

$$2\,\mathrm{Re}(g_0 z) - \int_{I_0}^{b_1(z)} dI \left( \left( I + \frac{I_0^2}{I} \right) \mathrm{Re}\,X_b \left( 0,\ I + \frac{I_0^2}{I} \right) \right)^{-1} = 2\,\mathrm{Re}(g_0 z_0) = \mathrm{const}. \quad (4.17)$$

The constants $I_0$ and $Z_0$ introduced here have a simple meaning. $I_0$ is an average intensity in the cross section $z = z_0$, where the colliding beams are equally intense. The function $X_g$ is determined by formula (3.1).

In Fig.4.4, the intensities $I_1, I_2$ of the colliding waves are presented as the function of the longitudinal coordinate $z$ to demonstrate the difference between the case of interaction of plane waves and the case, when the colliding waves have a complicated structure. Since systems (4.15) and (4.16) are translationally invariant, the point of reference can be placed anywhere. The only selected point is the cross section $z = z_0$, where the intensities of the colliding waves coinside. The pattern is symmetric with respect to the ordinate axis, therefore, only one half of the graph is presented in Fig.4.4. The solid curves and the dashed curves relate to the cases of the plane waves and the modulated waves, respectively. For each of these cases, several pairs of curves corresponding to boundary conditions $I_1(0) = I_2(0) = 0.2$, $I_1(0) = I_2(0) = 1$, $I_1(0) = I_2(0) = 6$ depart from the straight line $z = 0$. The lower curve of each pair corresponds to the "second" wave that propagates to the left in the process of amplification. The upper curve corresponds to the "first" wave that propagates to the right. Such curves for plane waves are given also in [81].

It can be noted that in the case of plane waves, the wave achieving the cross section $z = 0$ with a fixed intensity $I_0$ could not be amplified from the intensity level as small as desired in a sufficiently long amplifier. Everywhere in the amplifier, it had intensity greater than $I_{min} = I_0 + 1/2 - (I_0^2 + 1/4)^{1/2}$. For modulated waves, the situation is different. For a given intensity in the cross section $z = 0$, the wave at the amplifier input may be weak, if the amplifier is sufficiently long.

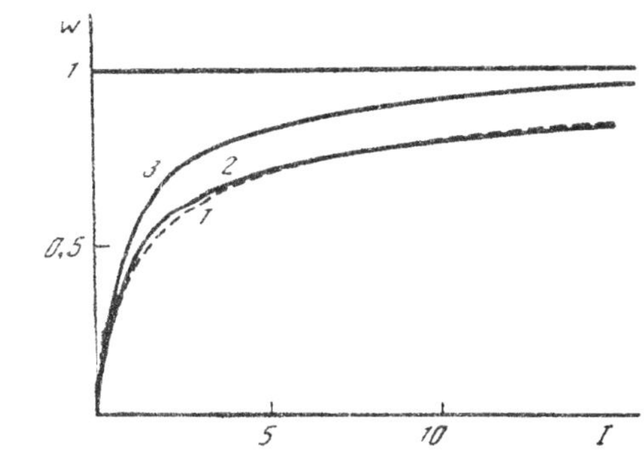


Fig. 4. 6. Dependence of a normalized amplification coefficient of a double-pass amplifier on the input intemsity at different values of amplification $g_oL : 2g_oL$ = 0 (1), 0. (2), 0.5 (3), 1 (4), 2 (5).

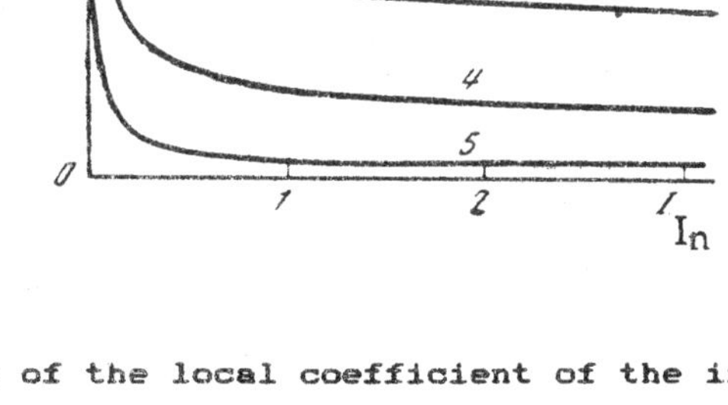


Fig. 4. 5. Dependence of the local coefficient of the inversed population utilization $w$ on the intensity $I = I_1 + I_2$.

1, the case of modulated waves (formula (4.19)); 2, the case of standing wave (formula (4.20); 3, the case of running wave (formula (4.1).

Let us consider the influence of the interaction described above upon the properties of an amplifier with colliding waves. To compare the cases of two plane waves and two randomly modulated waves, we introduce a quantity characterizing utilization of the active medium, i.e. the sum of products of intensities of the components under consideration and the corresponding increments:

$$u = 2\,\mathrm{Re}\,(X_{a1}a_1 + X_{b1}b_1 + X_{a2}a_2 + X_{b2}b_2). \tag{4.18}$$

Power extracted from the amplifier is equal to integral of $u$ over the volume. The quantity $w = u/\mathrm{Re}(2g(0))$ may be called efficiency of inversed population utilization. For plane waves, this qiantity is

$$w_p = 1 - (1 + 2I_1 + 2I_2 + (I_1 - I_2)^2)^{-1/2}, \tag{4.19}$$

and for completely modulated waves it is

$$w_m = (I_1 + I_2)\,X_b(0,\ I_1 + I_2)/\mathrm{Re}\,(2g_0). \tag{4.20}$$

In formulae (4.19) and (4.20), $I_1$ and $I_2$ are the intensities of the colliding waves. We emphasize that for modulated waves, $w$ depends on only one parameter (total intensity); for the plane waves, $w$ is determined by two parameters.

Dependence of the efficiency $w$ on the total intensity $I = I_1 + I_2$ is presented in Fig. 4.5. Curve 1 relates to the case of completely modulated waves and describes dependence of $w$. Curve 2 describing dependence of efficiency $w$ for colliding plane waves of equal intensities

$$w_s = w_p|_{I_1+I_2=I,\ I_1-I_2=0} = 1 - (1 + 2I)^{-1/2}. \tag{4.21}$$

is rather close to the first curve. For the intensities $I < 4.6$, the standing wave takes somewhat more energy from a unit volume, and for $I > 4.6$, the modulated wave is somewhat more "productive", the relative difference between the curves 1 and 2 in Fig.4.5 does not exceed 6% though. For comparison, dependence of the coefficient $w_{2}$ for the running wave on the intensity is also presented in Fig.4.5 (curve 3):

$$w_{\eta} = w_{p}\big|_{|I_1 - I_2| = I_1 + I_2 = I} = 1 - 1/(1 + I). \tag{4.22}$$

Let us apply the obtained results to the calculation of a double-pass amplifier (Fif.4.2,b). Let after a single pass through the amplifier, an optical wave with a random spatial modulation strike a mirror. After the second passage through the amplifier, the beam leaves the sustem freely. In this case, colliding waves with different transverse structure propagate in the amplifying medium, and they are described by system (4.16). Dependence of amplification coefficient $K = I_{out} / I_{in}$ normalized by the small-signal amplification coefficient $K_0$ on the input intensity $I_{in}$ is shown in Fig.4.6. Analogous curves for the plane waves are quite close to the curves for the modulated waves, though the intensities on the mirror differ noticeably in these two cases.

Note that the coefficients $A_1$, $C_1$, $A_2$, $C_2$ were assumed to be equal to zero on the whole nonlinear medium length $L$, when system (4.10) was solved for the case of completely modulated waves. This is possible only in the absence of mutually conjugated components in the colliding beams. Their presence results in degenerate four-wave interaction. One of these cases is considered below.

## 4.4. Contrast reversal in an extended amplifier

When colliding partially conjugated beams propagate in an extended nonlinear amplifier, the plane components of these beams act as reference waves. In this case, effects appear analogous to degenerate phase conjugation. These effects can manifest themselves in both image amplification [42-51] and active formation of a field with a definite spatial structure [55-60]. In this section, interaction of two colliding modulated mutually conjugated waves on the background of plane waves in an amplifying medium with gain saturation is investigated.

Interaction of conjugated waves on the background of plane waves corresponds, in particular, to the following combination of coefficients (see 4.9)

$$A_1 \neq 0, \quad A_2 \neq 0, \quad B_1 \neq 0, \quad C_2 \neq 0, \quad C_1 = B_2 = 0. \tag{4.32}$$

Direct test shows that these equations result in $\rho_{b2} = \rho_{c2} = 0$, i.e. conditions (4.23) are self-consistent. Besides, system (4.10) corresponds to the situation in which a plane wave with an amplitude $A_1(0)$ and a completely modulated component $f_2$ with an amplitude $B_1(0)$ are introduced into an extended amplifier from one end, and only a plane wave with an amplitude $A_2(L)$ or a certain linear combination of a conjugated field $f_1^*$ with this plane wave is introduced from the other end. This may be realized in experiments according to the layout shown in Fig. 4.2, if one of the transparencies is removed, and also in Fig. 4.2, b, if a mirror (element 5) is replaced by a phase conjugating system.

First we consider the case, when the coefficients with subscript 2 are essentially smaller in modulus than the coefficients with subscript 1 (the radiation propagates primarily along $z$ axis, and the counter-propagating beam is weak). This situation takes place in amplification of a reversed wave formed by a PC mirror with a small coefficient of reversal.

Expanding the function $\mathcal{F}$ in $|E_2|$ in formulae (4.10) and restricting ourselves to the linear terms, we obtain a separable system

$$dA_1/dz = X_a(|A_1|^2, |B_1|^2)A_1, \quad dB_1/dz = X_b(|A_1|^2, |B_1|^2)B_1 \tag{4.24}$$

for the field $E_1$ (it coinsides with (2.8)) and a linearized system

$$\begin{aligned} -dA_2/d(g_0 z) &= S_{11}A_2 + S_{12}C_2, \\ -dC_2/d(g_0 z) &= S_{21}A_2 + S_{22}C_2. \end{aligned} \tag{4.25}$$

for the field $E_2$ propagating in the opposite direction. For a purely active nonlinearity of saturation type ($g_0 > 0$, $I_s = 1$), the coefficients $S$ may be expressed through the intensities $a_1$, $b_1$ as

$$S_{11} = \int_0^\infty r\,dv, \qquad S_{12} = -\int_0^\infty 2\sqrt{a_1/b_1}\,(v/b_1)\,r\,dv,$$

$$S_{21} = S_{12}, \qquad S_{22} = \int_0^\infty (v/b_1)\,r\,dv = X_b(a_1, b_1), \tag{4.26}$$

where $r = (1 + a + v)(1 + 2a_1 + 2v + (a_1 - v)^2)^{3/2}\exp(-v/b_1)/b_1$. For clearness, these coefficients are presented in Fig. 4.7 by the level lines as the functions of the intensities $a_1$, $b_1$. The mixed coefficient $S_{12}$ is negative, a weak field is amplified more efficiently at $|\mathrm{Arg}(C_2/A_2)| \approx \pi$ than at $\mathrm{Arg}(C_2/A_2) = 0$. Let us examine (3.25) at fixed values of the coefficients $S$ to illustrate this fact. In this (somewhat artificial) case, the fundamental solution of system (4.25) has the form of

$$A_{2\pm} = \exp(\lambda_\pm g_0 z), \qquad C_{2\pm} = -\frac{\lambda_\pm - S_{11}}{S_{12}}\exp(\lambda_\pm g_0 z), \tag{4.27}$$

where

$$\lambda_\pm = \pm((S_{11} - S_{22})^2/2 + S_{12})^{1/2} - (S_{11} + S_{12})/2. \tag{4.28}$$

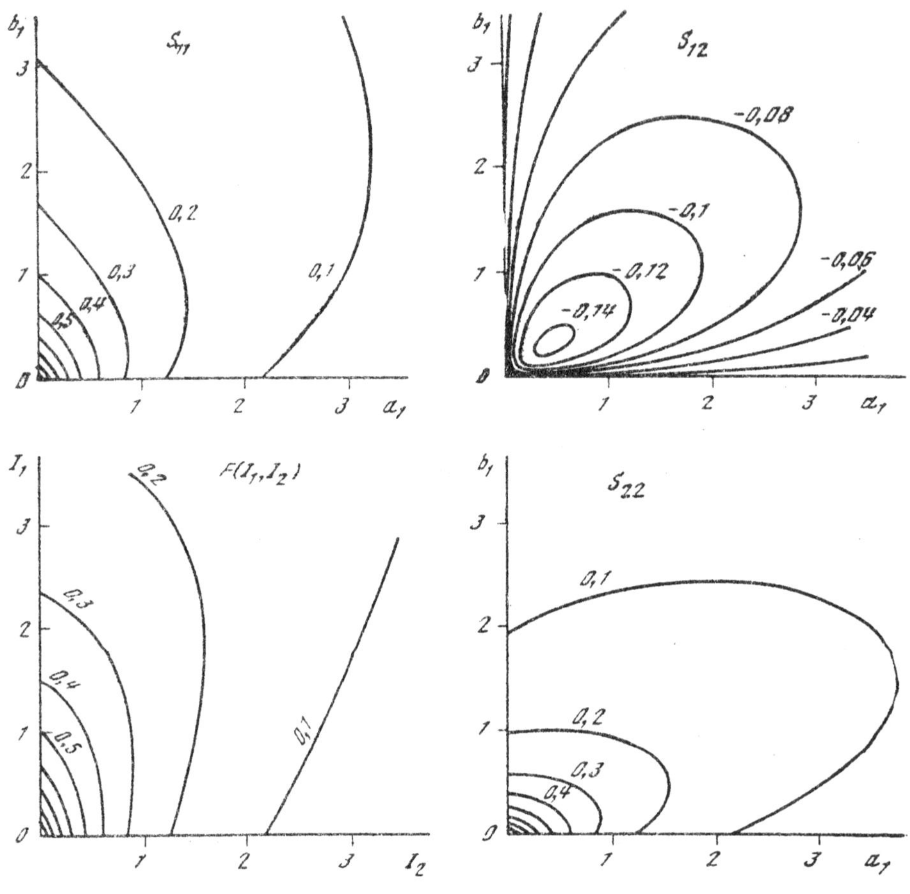

Fig. 4.7. Level lines of the coefficients $S$ determining amplification of a weak partially modulated field on the background of a strong counter-propagating field calculated in accordance with formulae (4.26); level lines of the function $\mathcal{F}(I_1, I_2)$ determining local interaction of the colliding beams (formula (1.10) at $I_s = 1$.

The sign before the radical in formula (4.27) determines the sign of the ratio $C_2/A_2$; $C_2/A_2 < 0$ for the minus sign and $C_2/A_2 > 0$ for the plus sign. In the first case, formula (4.27) corresponds to a negative (intensity maxima of the field $A_2 + C_2 f_1^*$ coinside with intensity minima of the field $A_1 + B_1 f_1$, see model example in Fig. 1.3). In the second case, the formula corresponds to a positive. It follows from (4.28) that the combination $A_2 + C_2 f_2$ negative with respect to the field is amplified better. If at the right end ($z = L$) only a plane wave is introduced into the amplifier ($A_2(L) > 0$, $C_2(L) = 0$), this field may be presented in the form of a linear combination of a positive and a negative. Preferential amplification of a negative corresponds to appearance of a conjugated field $f_1^*$ with a negative coefficient. In this sense, one may consider that in the presence of degenerate four-wave phase conjugation in a resonantly amplifying medium, the conjugated field carries an image negative with respect to that carried by the initial field together with the copropagating reference wave. In the case of resonantly absorbing medium ($g_0 < 0$), $C_2 > 0$, and the reversed wave carries a positive.

Formulae (4.27), (4.28) qualitatively illustrate the formation of a negative image in the amplifier. They do not take into account variations of the coefficients $S$ with the intensities $a_1, b_1$ growing. Amplification of the "first" field limits the region of efficient interaction of the "second" field components. The range of values of $a \approx b \approx 0.5$, in which $S_{12}$ differs noticeably from zero, can be seen from Fig. 4.7.

In system (4.24), (4.25), the first pair of equations is independent, its solutions were considered in [39]. One of them corresponding to the boundary conditions $A_1(0) = B_1(0) = \sqrt{0.05}$ is repeated in the upper graph of Fig. 4.8. Solutions of system (4.25) corresponding to the

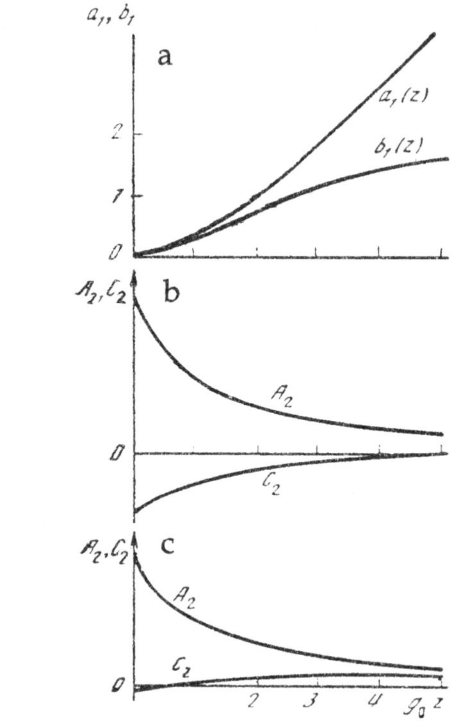


Fig. 4.8. Growth of the intensities $a_1, b_1$ (a) in the amplifier (solutions of system (4.4) with the boundary condition $a_1(0) = b_1(0) = 0.05$); variation of the coefficients $A_2$ and $C_2$ at $A_2 > 0$, $C_2(L) = 0$ (b) and at $A_2(L) = 2C_2(L) > 0$ (c).

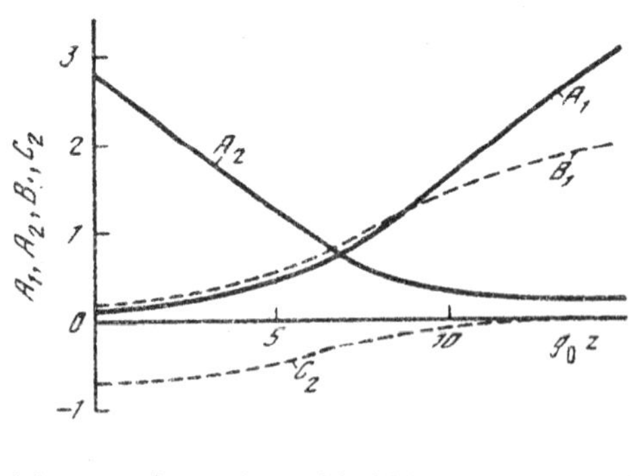


Fig. 4.9. One of solutions of system (4.10) corresponding to the boundary conditions $A_1(0) = 0.10$; $B_1(0) = 0.12$; $C_1(0) = 0$; $A_2(L) = 0.24$; $B_2(L) = 0$; $C_2(L) = 0$.

boundary conditions $A_2(L) > 0$, $C_2(L) = 0$ and $A_2(L) = 2C_2(L) > 0$ at $g_0L = 5$ are depicted lower. Fig. 4.8 shows that a plane wave entering the amplifier gives birth to a conjugated wave $f_1^*$ carrying a negative image, and if a positive image is introduced to the amplifier input $z = L$, interaction of this wave with the strong field $E_1$ results in the modulation depth decrease down to zero, and then the image transforms into a negative (the lower graph).

Limitations of the region of efficient interaction of the field $E_2$ components may be associated not only with nonlinearity saturation by the field $E_1$, but also with saturating action of the field $E_2$ itself following its amplification. Eqs. (3.24), (3.25) are suitable only for small values of $a_2, c_2$ (as $z \to -\infty$, they give exponential growth of the coefficients $A_2, C_2$). If $E_1$ is comparable with $E_2$, system (4.12) should be employed. One of realizations of dependence of the amplitudes $A_1$, $B_1$, $A_2, C_2$ on $g_0z$ is presented in Fig. 4.9. It corresponds to appearance and amplification of a conjugated wave $f_1^*$ carrying a negative image together with a reference plane wave $\mathcal{A}_2$. Note that in the right part of the graph, linearized system (4.24), (4.25) is valid, in the left part, the same system may be used with $z \to -z$, "1"$\leftrightarrow$"2", $B \leftrightarrow C$. In both right and left parts of the graph, the modulated component tends to a certain constant value [39].

Note that contrast reversal in the counter-propagating field is associated with the presence of definite phase relations between the reversed wave and the reference one. Therefore, if the reference wave is separated from the reversed wave, as it is implied in the standard configuration of four-wave PC (see, for instance, [27]), it is not possible to speak of a positive or a negative in the reversed wave. However, in the systems for image amplification, there is no separation of the reference wave from the reversed wave. So contrast reversal in amplifiers is quite observable (see, for instance, [49]).

Analogous contrast reversal in refractive media was discussed in [29,75].

## 4.5. Medium with pseudolocal response

In accordance with the stated problem, in previous sections of this chapter, propagation of radiation in a medium with local response was considered. However, in some cases a local increment is not determined by a modulus squared of the field in a single point. The case in which an interference pattern formed by colliding beams is completely washed out, and the grating induced in the medium by each of these beabs is not washed out at all, has a comparatively simple interpretation. In this case, the increment may be considered to depend locally on the total intensity of the colliding beams (the same situation occurs in a medium with local response, if the colliding beams are polarized in mutually perpendicular directions). This is the reason for the choice of the title of this section.

In analogy with Sec.1.1, let us examine approximate solution of Eq. (1.1) of the form of (1.3) where the colliding waves are separated explicitly. Assuming that on a scale of the order of $\lambda/2$ the increment $g$ varies insignificantly and pulling it outside the averagings over $z$, we come to system (1.8) with the only difference that the function $F$ is described not by formula (1.7) but by formula

$$F\,(I_1,\ I_2) = g\,(I_1 + I_2). \tag{4.29}$$

Note that in this formula, intensities rather than amplitudes of the colliding waves are added. Dependence of the form of (4.29) was used in [83,84]. Analysis performed in the previous sections without utilization of explicit form of the function $F(I_1, I_2)$ remains valid. Therefore, to find out how sensitive the conclusions of the previous

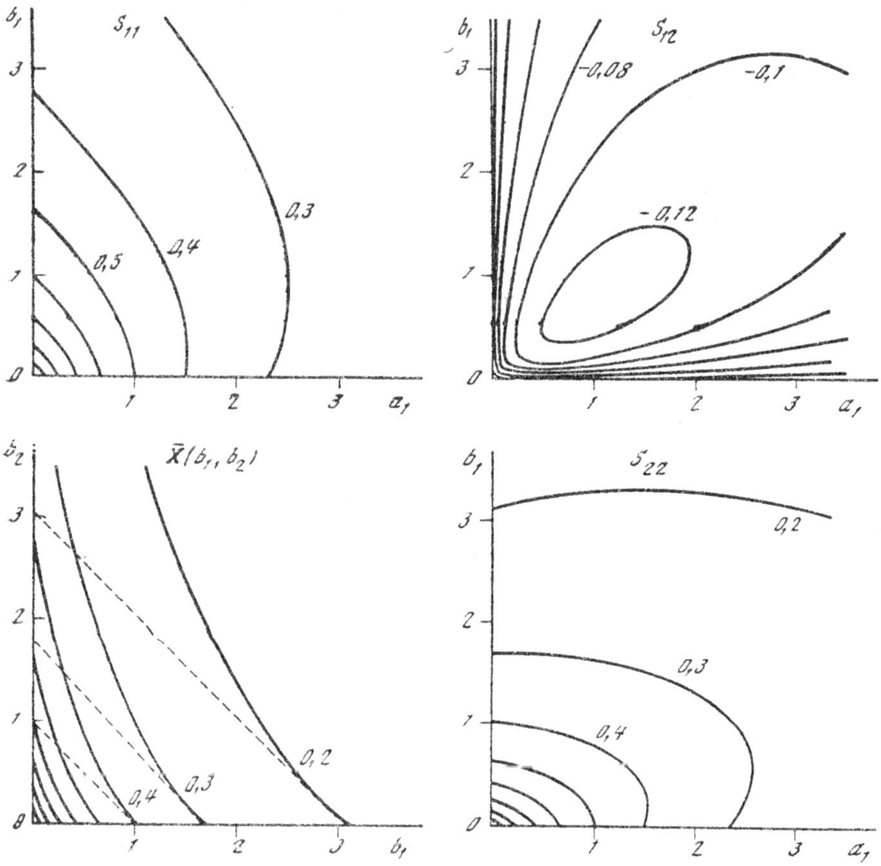


Fig. 4.10. Level lines of the effective increment $\bar{X}(b_1, b_2)$ of the first beam calculated in accordance with formula (4.31) for completely modulated colliding waves in the case of pseudolocal response (lower left); level lines of the coefficients $S$ of linearized system (4.6) calculated in accordance with formulae (4.32) for this case.

sections are to locality of medium response, only particular cases should be considered.

Let us consider interaction between colliding completely modulated waves in a medium with pseudolocal response. Repeating derivations from Sec. 4.3 for the case of dependence of the form of (4.29), we get

$$db_1/dz = 2\overline{X}(b_1,\ b_2)\, b_1, \qquad db_2/dz = -2\overline{X}(b_2,\ b_1)\, b_2, \tag{4.30}$$

where

$$\overline{X}(u,\ v) = \int_0^\infty\!\!\int (\exp(-\xi-\eta))\, g(u\xi + v\eta)\, \xi\, d\xi\, d\eta. \tag{4.31}$$

Function $\overline{X}$ of the form of (4.31) is presented in Fig. 4.10 (the left lower graph). It is seen from the figure that in the case of waves with a complete spatial modulation, amplification of each beam is saturated by the beam itself more efficiently that by the counter-propagating beam (it was shown in Secs. 1.2 and 4.3 that in the case of plane waves in a medium with local response, the situation is opposite, i.e. the effective increment of the weak wave is smaller than that of the strong one). The case of medium with local response, when the effective increments of the colliding waves coinside and depend only on the total intensity (formula (4.15)), is presented by the dashed curves in the same figure, for comparison. The difference is noticeable only for a weak beam on the background of a strong one.

To find out how strongly the conclusions about the formation of a negative image in amplification of colliding waves with a spatial modulation depend on locality of the medium response, let us examine system (4.30) for a weak beam on the background of a strong one in a medium with pseudolocal response. For a function of the form of (4.29), the coefficients $S$ are

$$S_{11} = \int_0^\infty r\,dv, \qquad S_{12} = S_{21} = \int_0^\infty \frac{r\,dv}{4a_1b_1}, \qquad S_{22} = \int_0^\infty vr\,dv, \tag{4.32}$$

where $r = (4a_1b_1)^{-1}/(1 + 2a_1 + b_1v + (a_1 - b_1v)^2)^{1/2}$. These functions are presented in the upper graphs and in the right lower graph of Fig. 4.10. The mixed coefficient $S_{12}$ is responsible for the formation of a negative. We note that in this case it is negative as well. It may be noted from comparison of Figs. 4.7 and 4.10, that extremum values of $S_{12}$ in these two cases are of the same order ($\min S_{12} \approx -0.1$), but in the case of pseudolocal amplification saturation, the extremum value is achieved at larger intensities ($a_1 \approx b_1 \approx 1$, instead of $a_1 \approx b_1 \approx 0.5$). The difference between these cases is not a qualitative one.

So, the conclusions of the previous section about formation of a negative image in a purely active medium with saturation should be applied also to the case, when the medium "senses" only intensity averaged over a period of interference pattern formed by colliding beams (as, for instance, in copper vapor amplifier).

## 4.6. Special features of joint amplification of colliding optical beams (conclusions of Ch. 4)

Nonlinear interaction of colliding optical beams depends on the presence of conjugated components in them. Therefore, if both plane mutually conjugated waves and partially conjugated modulated components are present in the colliding beams, their interaction in a weakly nonlinear medium is described by a system of six ordinary differential equations for the coefficients in the field expansion (4.9) in these components. Independent beams with complete spatial modulation in medium with local response are amplified with equal increments. This case differs essentially from the case of colliding plane wave amplification, when the increment of the strong wave is greater than that of the weak one.

In propagation of colliding partially modulated waves, their plane components act as reference waves. It leads to the effects analogous to four-wave phase conjugation. In an amplifying medium with saturation, degenerate four-wave interaction results in formation of a reversed wave, its phase is related in a definite way to the phases of the reference waves. Since in this case intensity maxima of the field $E_1$ coinside with intensity minima of the field $E_2$ , this relation may be interpreted as a contrast reversal. In this sense, it is possible to say that the reversed wave carries a negative image.

This effect is essential in joint nonlinear amplification of a beam with partial spatial modulation and a conjugated field formed outside the amplifier. In the conjugated beam, the modulation depth diminishes more rapidly than the analysis of a single partially modulated beam predicts. Fast drop of a modulation depth (down to contrast reversal) can influence radically the operation properties of an amplifier with a PC mirror. In the reversed beam, the informative part (the modulated component) may be lost.

Although phase conjugation in degenerate four-wave interaction in nonlinear media has been investigated for a long time and was studied thoroughly enough (see, for example, [27,28,30,31], the relations between the phases of the reversed wave and the reference wave has been discussed theoretically [29] and investigated experimentally [77] only recently. Contrast reversal in four-wave mixing in nonlinear media is quite a general phenomenon. Since the image contrast is an essential characteristic of optical signals, the relations between the phase of the reversed wave and the phases of the reference beams is of great practical importance.

## Conclusions

It was shown that in actual amplifiers the influence of the refractive nonlinearity upon the radiation structure may be as essential as that of the amplification saturation. Let us summarize the basic conclusions of the present paper.

1. In an extended amplifier, transformation of a wide beam with a complex structure with partial spatial modulation is described by a system of two ordinary differential equations in terms of interaction between the completely modulated field component and the plane wave. Validity of such description is supported by independent investigation of variation of integral quantities determined by the Noether theorem and also by direct numerical computations.

2. If the amplification saturation results in a less efficient amplification of the modulated field component in comparison with the plane wave, a phase shift appears between these two components in the presence of nonlinear refraction. The second effect as well as the first one decreases the contrast of an image carried by the field; at large dephasing values, contrast reversal is obtained. Sharp dependence of the image contrast on the dephasing provides a technique for measuring the nonlinear refraction constant.

3. In the nonlinear amplification mode, an image is distorted by noise. The relative intensity of the latter is equal in order of magnitude to the product of the increment modulus squared, the amplifier length and the field diffraction length. Besides, in a self-focusing medium, an image is destroyed due to the plane wave instability at large dephasing values.

4. Effective increments of colliding completely modulated beam in an extended amplifier with local nonlinearity are identical. In the case of complete modulation of the colliding beams, degenerate four-

wave interaction leads to formation of reversed waves carrying negative images. This effect is particularly significant in nonlinear amplification of a partially modulated field and a reversed wave formed outside the amplifier. At large amplification values, the modulation depth in the reversed wave decreases down to contrast reversal.

The results of the analysis proposed may be utilized in designing the systems of optical information processing, applications of image amplifiers, and development of systems for active formation of an optical field with a predetermined structure.

## References

1. Akhmanov S.A., Khokhlov R.V. "Problems of nonlinear optics" Moscow: Nauka, 1964, 295 pp.
2. Akhmanov S.A., Sukhorukov A.P., Khokhlov R.V. "Self-focusing and diffraction of light in a nonlinear medium" UFN, 1967, vol.93, pp. 19-70.
3. Bespalov V.I., Litvak A.G., Talanov V.I. "Self-action of electromagnetic waves in third-order isotropic media" Nonlinear optics. Novosibirsk: Nauka, 1968, pp. 428-463. (Proc. 2-nd All-Union symp. on nonlinear optics.
4. Vinogradova M.B., Rudenko O.V., Sukhorukov A.P. "Theory of waves" Moscow: Nauka, 1979, 383 pp.
5. Kuznetsova T.I. "On interaction of electromagnetic waves with active media" Trudy FIAN, 1960, vol.43, pp. 146-160.
6. Kogelnik H. "On the propagation of Gaussian beams of light through lenslike media including those with a loss or gain variation" Appl.Opt., 1965, vol.4, pp.1562-1577.
7. Hardy A. "Gaussian modes of resonators containing saturable

gain medium" Ibid., 1980, vol.19, pp.3830-3836.

8. Vlasov S.N., Petrishchev V.A., Talanov V.I. "Averaged description of wave packets in linear and nonlinear media" Izv.vuzov.Radiofizika, 1971, vol.19, pp.1353-1363.

9. Lam J.F., Lippmann B., Tappert F. "Moment theory of self-trapped laser beams with nonlinear saturation" Opt.Commun., 1975, vol.15, pp.419-421.

10. Sydam B.R. "Self-focusing of very powerful laser beams. 2" IEEE J.Quant.Electron., 1974, vol.10, pp.837-843.

11. Dyshko A.L., Lugovoy V.N., Prokhorov A.M. "Self-focusing of intense light beams" Pis'ma v ZhETF, 1967, vol.6, pp.655-659.

12. Alonso L.M. "The nonlinear Schroedinger equation as a Galilean invariant system" J.Math.Phys., 1982, vol.23, pp. 1518-1523.

13. Garmire E., Chiao R.Y., Townes C.H. "Dynamics and characteristics of the self-trapping of intense light beams" Phys.Rev.Lett., vol.16, pp.347-349.

14. Yanauskas Z.K." Radial field distribution in in a self-focused light beam" Izv.vuzov.Radiofizika, 1966, vol.19, pp.412-415.

15. Yanauskas Z.K. "Dependence of a self-focused light width on intensity" Pis'ma v ZhETF, 1967, vol.5, pp.335-336.

16. Bogolyubov N.N., Shirkov D.V. "Introduction into the theory of quantized fields" Moscow: Nauka, 1984, 597 pp.

17. Itsikson K., Zuber J.B. "Quantum field theory" Moscow: Mir, 1984, vol.1, 488 pp.

18. Kuznetsov E.A., Turitsyn S.K. "Talanov transformation in self-focusing problem and instability of stationary waveguides" Phys.Lett.A, 1985, vol.112, pp.273-275.

19. Siegman A.E., Shiklas E.M. "Mode calculations in unstable resonators with flowing saturable gain. 1. Hermite-Gaussian

expansions" Appl.Opt., 1974, vol.13, pp.2775-2792.

20. Shiklas E.M., Siegman A.E. "Mode calculations in unstable resonators with folwing saturable gain. 2. Fast Fourier transform method" Ibid., 1975, vol.14, pp.1583-1589.

21. Kandidov V.P. "Investigation of optical field statistics using Monte-Carlo method" Izv.vuzov.Fizika, 1983, vol.47, pp.1583-1689.

22. Kandidov V.P., Lednev V.I. "Investigation of thermal self-action of a light pulse in a turbulent atmosphere using statistical test method" Kvantovaza elektronika, 1984, vol.8, pp.873-877.

23. Vorob'ev V.V. "Thermal self-action of laser radiation in atmosphere: theory and model experiment" Moscow: Nauka, 1987, 200 pp.

24. Zakharov V.E., Manakov S.V., Novikov S.P., Pitayevsky L.P. "Theory of solitons: inverse problem method" Moscow: Nauka, 1980, 380 pp.

25. Akhmanov S.A., Dyakov Yu.E., Chirkin A.S. "Introduction into statistical radiophysics and optics" Moscow: Nauka, 1981, 640 pp.

26. Pasmanik G.A. "Self-action of incoherent light beams" ZhETF, 1974, vol.66, pp.490-500.

27. Zel'dovich B.Ya., Pilipetsky N.F., Shkunov V.V. "Wavefront reversal" Moscow: Nauka, 1985, 248 pp.

28. Lazaruk A.M., Rubanov A.S. "Energy efficiency of wavefront reversal in induced and four-wave parametric scattering" Kvantovaya elektronika, 1980, vol.7, pp.1992-1995.

29. Lazaruk A.M. "Contrast inversion in a conjugate image reconstructed in four-wave mixing in media with reactive nonlinearity" Ibid., 1987, vol.14, pp.418-420.

30. Ivakin E.V., Koptev V.G., Lazaruk A.M. et al "Wavefront reversal of a light field in nonlinear interaction in saturable media" Pis'ma v ZhETF, 1979, vol.30, pp.648-651.

31. Hellwarth R.W. "Generation of time-reversed fronts by nonlinear refraction" J.Opt.Soc.Amer., 1977, vol.7, pp.1-3.

32. Zel'dovich B.Za., Yakovleva T.V. "Distortions of a fine-structure wave pattern by self-focusing nonlinearity" Kvantovaya elektronika, 1980, vol.7, pp.1325-1327.

33. Blashchuk V.N., Zel'dovich B.Ya, Shkunov V.V. "Four-wave wavefront reversal in the field of coded reference waves" Ibid., pp. 2559-2566.

34. Zel'dovich B.Ya., Lerner P.B. "Energy exchange between waves in recording and reconstruction in holographic media with regeneration response" Ibid., 1983, vol.10, pp.1764-1770.

35. Zel'dovich B.Ya., Shkunov V.V. "Specklon" Izv.AN SSSR. Ser.fiz., 1984, vol.48, pp.1545-1563.

36. Zel'dovich B.Ya., Shkunov V.V., Yakovleva T.V. "Noise calculation of volume holograms of speckle-fields" UFN, vol.149, pp.511-549.

37. Rytov S.M., Kravtsov V.I., Tatarsky V.I. "Introduction into statistical radiophysics" Moscow: Nauka, 1978, part 2, 349 pp.

38. Gochelashvili K.S., Chashey I.V., Shishov V.I. "Light pulse instability in a nonlonear scattering medium" Kvantovaya elektronika, 1980, vol.7, pp.2077-2082.

39. Kuznetsova T.I., Kuznetsov D.Yu. "Theoretical analysis of image transmission through an amplifying medium" Trudy FIAN, 1991, vol.206, pp.101-115.

40. Basov N.G., Danilov A.E., Dunnebier G. et al "System for radiation concentration of mutlichannel laser installation "Delphin-1" for heating and compression of thermonuclear targets in experiments on laser-produced thermonuclear fusion" Ibid., 1985, vol.149, pp.3-21.

41. Aleksandrova I.V., Basov N.G., Danilov A.E. et al "Limiting possibilities of laser heating of spherical targets" Ibid., pp. 42-59.

42. Zemskov K.I., Isayev A.A., Kazaryan M.A., Petrash G.G. "Laser projection microscope" Kvantovaya electronika, 1974, vol.1, pp. 14-15.

43. Zemskov K.I., Isaev A.A., Kazaryan M.A., et al. "Active optical system with brightness amplification by pulsed metal vapor lasers" Opt.Commun., 1976, vol.18, p.144.

44. Zemskov K.I., Isayev A.A., Kazaryan M.A., Petrash G.G. "Investigation of basic characteristics of laser projection microscope" Kvantovaya elektronika, 1976, vol.3, pp.35-43.

45. Doroshkin A.A., Zemskov K.I., Isayev A.A., et al."Dynamic information imaging on large screens using brightness amplifiers" Ibid., pp.2515-2516.

46. Zemskov K.I., Kazaryan M.A., Pekhoshkina T.I., Trofimov A.N. "Projection system with brightness amplifier based on copper chloride vapors" Ibid., 1979, vol.6, pp.391-394.

47. Zemskov K.I., Kazaryan M.A., Savransky V.V., Shafeyev G.A. "Laser projection microscope in transmitted light" Ibid., pp. 2473-2475.

48. Zemskov K.I., Kazaryan M.A. "Image brightness amplifiers based on lead and manganese vapors" Kratkiye soobshch. po fizike FIAN, 1979, No.5, pp.34-37.

49. Bunkin F.V., Zemskov K.I., Kazaryan M.A., et al. "Power self-regulation and negative image formation in illuminating beam of a laser projection microscope" Kvantovaya ekektronika, 1981, vol.8, pp.1372-1373.

50. Kazaryan M.A., Matveyev V.M., Petrash G.G. "Projection system with brightness amplifier and autonomous illumination source" Izv.AN SSSR, ser.fiz, 1982, vol.46, pp.1898-1904.

51. Zemskov K.I., Kazaryan M.A., Matveyev V.M., Petrash G.G. "Image contrast in laser projection microscope system" Kvantovaya elektronika, 1983, vol.10, pp.336-341.

52. Dumarevsky Yu.D., Zemskov K.I., Kazaryan M.A., et al. "Image projection on a large screen using MDS-LC structures and quantum

amplifiers" DAN SSSR, 1987, vol.292, pp.604-607.

53. Zemskov K.I., Kazaryan M.A., Matveyev V.M., et al. "Laser processing of objects with simultaneous visual control in the system of copper vapor laser oscillator and amplifier" Kvantovaya elektronika, 1984, vol.11, p.418.

54. Hansh T.F., Varsanyi F., Shawlow A.L. "Image amplification by dye lasers" Appl.Phys.Lett., 1971, vol.18, pp.108-110.

55. Chernega N.V., Kudryavtseva A.D., Sokolovskaya A.I. "Amplitude-phase screen in media with two-photon light absorption" In: Application of holography methods in science and technology. Leningrad: Nauka, 1987, pp.48-51. (Proc.18-th school on holography. Chernogolovka, 1987).

56. Kruzhilin Yu.I. "Self-aligning system "laser-target" for laser-produced thermonuclear fusion" Kvantovaya elektronika, 1978, vol.5, pp.625-631.

57. Hardy V. "Active image formation in lasers" In: Optical information processing. Moscow: Mir, 1966, pp.334-337.

58. Kuznetsova T.I., Kuznetsov D.Yu. "On reconstruction of optical field phase characteristics using active optical systems" Kvantovaya elektronika, 1985, vol.12, pp.2507-2509.

59. Kuznetsova T.I., Kuznetsov D.Yu. "On reconstruction of complex light field on the basis of optical processing of amplitude transparencies" Kratkiye soobshch. po fizike FIAN, 1986, No.2, pp.12-14.

60. Kuznetsova T.I., Kuznetsov D.Yu. "A phase retrieval algorythm based on analogy for optical image processing" Opt.Commun., 1987, vol.61, pp.374-376.

61. Kuznetsova T.I., Kuznetsov D.Yu. "Interaction of a spatially modulated wave of complex structure with a plane wave in a quantum amplifier" Kvantovaya elektronika, 1981, vol.8, pp. 1808-1815.

62. Kuznetsov D.Yu. "On interaction of a spatially modulated wave of complex structure with a plane wave in a medium with non-linear refraction" Ibid., 1986, vol.13, pp.630-632.

63. Kuznetsov D.Yu. "Measurement of nonlinear refraction constant with the help of monochromatic spatially modulated wave" Kratkiye soobshch. po fizike FIAN, 1986, No.2, pp.9-11.

64. Kuznetsov D.Yu. "On validity of the statistical hypothesis in description of propagation of monochromatic spatially modulated wave of complex structure in a weakly nonlinear medium" Moscow, 1986, 8 pp. (Preprint FIAN, No.42).

65. Kuznetsov D.Yu. "Influençe of saturable amplification on a self-focusing instability of plane wave" Kratkiye soobshch. po fizike FIAN, 1987, No.6, pp.15-17.

66. Kuznetsov D.Yu. "Lagrange formalism in application to active optical system" Ibid., No.1, pp.24-26.

67. Kuznetsova T.I., Kuznetsov D.Yu. "On interaction of colliding spatially modulated beams in nonlinear medium" Kvantovaya elektronika, 1984, vol.11, pp.2210-2215.

68. Kuznetsova T.I., Kuznetsov D.Yu. "Conditions of negative image appearance in optical signal amplifier" Ibid., pp.2185-2190.

69. Kuznetsova T.I., Kuznetsov D.Yu. "Features of four-wave wavefront reversal in amplifying medium" In: Laser radiation wavefront reversal in nonlinear media. Minsk. Physics institute of AN BSSR, 1987, pp.73-89.

70. Born M., Wolf E. "Principles of optics" Moscow: Nauka, 1970, 865 pp.

71. Lifshits E.M., Pitayevsky L.P. "Electrodynamics of continuous media" Moscow: Nauka, 1982, 620 pp.

72. Moran M.J., She C.Y., Carman R.L. "Interferometric measurements of the nonlinear refractive-index coefficient relative to CS

in laser-system-related materials" IEEE J.Quant.Electron. 1975, vol.11, pp.259-263.

73. Vasil'eva M.A., Vishchakas Yu., Gul'binas V., et al. "Measurements of nonlinear refractive index of laser active media with Nd " Kvantovaya elektronika, 1985, vol.12, pp.997-1001.

74. Karlov N.V., Karpov N.A., Petrov Yu.N., Stel'makh O.M. "Self-focusing of $CO_2$ -laser radiation in resonant absorbing media" Pis'ma v ZhETF, 1973, vol.17, pp.337-340.

75. Ben' V.N., Ivakin E.V., Rubanov A.S. "Contrast inversion in reversed wave in degenerate four-wave interaction by pump phase modulation" In: Laser radiation wavefront reversal in nonlinear media. Minsk: Physics institute of AN BSSR, 1987, pp.131-138.

76. Isayev A.A. "Spectral content of induced radiation of a copper vapor pulsed laser" Kvantovaya elektronika, 1980, vol.7, pp. 599-609.

77. Vasil'eva M.A., Gul'binas V., Kabelka V., et al. "Dispersion of dye solution phase response following picosecond excitation" Ibid., 1984, vol.11, pp.1431-1436.

78. Vishchakas Yu., Gul'binas V., Kabelka V. "Measurement of dispersion of R6G and DODCI dye solution phase response" Ibid. 1985, vol.12, pp.1989-1990.

79. Kabanov V.V., Rubanov A.S. "Spectral dependence of refractive index in the region of absorption and amplification bands of organic dyes" Doklady AN BSSR, 1980, vol.24, pp.34-37.

80. Leontovich M.A., Fok V.A. "Solution of problem of electromagnetic wave propagation along the Earth surface using the parabolic equation method" ZhETF, 1946, vol.16, pp.557-573.

81. Kuznetsova T.I., Rautian S.G. "Plane solution of wave equation for a layer with a negative absorption coefficient with allowance for saturation" Izv.vuzov.Radiofizika, 1964, vol.7, pp.682-692.

82. Abrams P.L., Lind R.C. "Degenerate four-wave mixing in absorbing media: Errata" Opt.Lett., 1978, vol.2, p.205.

83. Abrams P.L., Lind R.C. "Degenerate four-wave mixing in absorbing media" Ibid., pp.94-96.

84. Bel'dyugin I.M., Zolotaryov M.V., Kireyev S.E., Odintsov A.I. "Cu vapor laser with self-pumped phase-conjugate mirror" Kvantovaya elektronika, 1986, vol.13, pp.825-827.

85. Ibragimov N.Kh. "Transformation groups in mathematical physics" Moscow: Nauka, 1983, 280 pp.

86. Kelley P.L. "Self-focusing of optical beams" Phys.Rev., 1965, vol.15, pp.1005-1008.

87. Elliot C.J. "Gain saturation and self-focusing consideration in the design of the optical amplifier" Appl.Phys.Lett., 1974, vol.24, pp.91-93.

88. Penzkoffer A., Wienhardt N. "Spectral development of picosecond pulses of mode-locked Nd-glass lasers" IEEE J.Quant.Electron., 1983, vol.19, pp.567-573.

89. Okulov A.Yu., Orayevsky A.N. "On compensation of self-focusing distortions in quasiresonant light amplification" Kvantovaya elektronika, 1987, vol.14. pp.369-375.

90. Vasil'eva M.A., Kazaryan M.A., Kuznetsova T.I., Petrash G.G. "On image distortion in quantum amplifiers" Ibid., 1978, vol.5, pp.666-669.

91. Gradshtein I.S., Ryzhik I.M. "Tables of integrals, sums, series, and products". Moscow: Nauka, 1971, 1108 pp.

92. Askaryan G.A., Diyanov Kh.A., Mukhametzhanov M. "Elimination of high-power beam self-collapse in a nonlinear medium using a raster; multiple waveguide energy propagation: Diffraction grating in a nonlinear medium" Pis'ma v ZhETF, 1972, vol.16, pp.211-215.

93. Zel'dovich B.Ya., Tabiryan N.V. "Orientational nonlinearity of liquid crystals" UFN, 1985, vol.147, pp.633-674.

94. Rabiner L., Gold B. "Theory and application of digital signal processing". Moscow: Mir, 1978, 848 pp.

95. Abramovits M., Stigan I. "Handbook on special functions". Moscow: Nauka, 1985, 832 pp.

96. Bel'dyugin I.M., Galushkin M.G., Zemskov E.M. "Propagation of spatially inhomogeneous radiation in nonlinear resonant media" Kvantovaya elektronika, 1983, vol.10, pp.2066-2070.

# THEORY OF DEGENERATE INTERACTION OF MULTIFREQUENCY IR RADIATION IN RESONANT MEDIA

*I.M. Bel'dyugin, A.A. Stepanov and V.A. Shcheglov*

## INTRODUCTION

Nonlinear multiwave processes realized by mixing two or more waves with equal frequencies in absorbing or amplifying resonant media are usually called degenerate resonant interactions. For example, interference interaction of counter-propagating waves in active laser media is a typical case of degenerate two-wave interaction (DTI) [1-3]. Similar to the previous, the mechanism of degenerate four-wave interaction (DFI) is widely used in the schemes of phase conjugation (PC) of IR radiation [4-7]. In principle, more complex (for example, six-wave and even eight-wave [8]) interactions are sometimes realized that are also related to degenerate interactions if the frequencies of interacting waves are equal.

It should be noted that the role of interference of counter-propagating waves in the amplification of monochromatic laser radiation in resonant media is often studied in the literature [3]. Recently, such investigations acquired a direct practical significance [9], in particular, in connection with the problem of devising high-power two-pass laser amplifiers with a low radiation divergence [10]. Nevertheless, it is essential that for a number of gas lasers that are practically used and generate multifrequency radiation on cascade transitions of two-atom molecules (CO-, HF-lasers, etc.), the methods previously developed in a single-frequency approximation and the equations obtained on their basis (describing two-wave interference interaction of counter-propagating waves) are not applicable in reality. Direct investigations show that a large number of frequencies introduce principally new features into the phenomenon under consideration.

Beginning with the paper by Abrams and Lind [11], many works were devoted to theoretical investigation of DFI in resonant media, but in most of publications the stationary theory was developed ([12-15] should be mentioned of the last works). The questions of nonstationary DFI in resonant media were studied in [16-19]. We note that these investigations were restricted mainly to the small-signal approximation that is not applicable, in particular, for the analysis of the so-called ring schemes [20-28] of self-conjugation of IR radiation based on DFI in the media with resonant and thermal nonlinearity.

The results given below were obtained in the Optics Laboratory of the Lebedev Physics Institute in investigation of the problem of degenerate multiwave processes taking place in resonant media with relaxation at arbitrary intensities of the interacting fields. Both cases of single-frequency and multifrequency IR radiation for cascade transitions in multilevel molecular systems are considered.

It is necessary to note that the problem of DFI in resonant media with relaxation in the fields with arbitrary intensity is analyzed in the recent publication [29] too. Using a somewhat different approach and neglecting the diffusive motion of the particles, a set of closed equations for DFI in the case of counter-propagating waves is obtained in it that describes the spatial and temporal evolution of the optical wave amplitudes without any limitation on the relation between the intensities of interacting waves. However, it is difficult to extend the approach used in [29] to the case of multifrequency DFI that is also analyzed in the present work.

# 1. TWO-WAVE (INTERFERENCE) INTERACTION OF IR RADIATION IN ABSORBING AND AMPLIFYING MEDIA

## 1.1. Single-frequency monochromatic radiation

With the aim of illustrating the method, we begin our examination addressing to the well-known case of interference interaction of two counter-propagating waves with the same frequency in a two-level absorbing (or amplifying) medium.

Restricting ourselves to homogeneous broadening of spectral lines, we introduce the absorption coefficient for a two-level resonant medium

$$g = \sigma (N_1 - N_2), \tag{1.1}$$

where $N_k$ is the population density of the $k$ th level ( $k = 1, 2$), $\sigma = \sigma_0/(1 + \delta^2)$ the frequency-dependent cross section of the radiative transition, $\sigma_0$ the cross section at the line center, $\delta = (\omega - \omega_0) T_2$ the frequency detuning, $T_2$ the polarization relaxation time, and $\omega_0$ the transition frequency at the line center.

Being interested mainly in molecular absorbing (amplifying) media, we shall suppose in the future that the characteristic time of population relaxation $T_1$ considerably exceeds the polarization relaxation time ( $T_1 \gg T_2$ ). In this case the temporal dynamics of absorption coefficient under the action of the radiation field is described by the obvious equation

$$\frac{\partial g}{\partial t} + \frac{1}{T_1}(g - \alpha_0) = -2\sigma g I, \tag{1.2}$$

where $\alpha_0$ is the unsaturated absorption coefficient in the absence of an external field; $I$ the total radiation intensity in the medium produced by counter-propagating waves.

To describe the interference interaction, we use the wave equation for the radiation field

$$\nabla^2 E = \frac{1}{c^2}\frac{\partial^2 E}{\partial t^2} + \frac{4\pi}{c^2}\frac{\partial^2 P}{\partial t^2}. \tag{1.3}$$

Here the electromagnetic field $E$ is the superposition of two counter-propagating waves ($E = E_1 + E_2$) and $P$ the polarization of the resonant medium.

In the case of monochromatic radiation we have $E = \mathcal{E}(\mathbf{r}, t)e^{i\omega t} +$ + c.c., where $\mathcal{E}(\mathbf{r}, t) = \mathcal{E}_1(\mathbf{r}, t) + \mathcal{E}_2(\mathbf{r}, t)$ is the slowly varying in time complex amplitude of the total radiation field. Setting down $P = \mathcal{P}(\mathbf{r}, t)e^{i\omega t}$ + c.c., like in the previous case, one can obtain the polarization of resonant media in the quasistatic approximation ($\partial\mathcal{P}/\partial t = 0$) using the well-known (see, for example, [30,31]) relation

$$\mathcal{P} = -\frac{i+\delta}{4\pi k} g\mathcal{E}, \tag{1.4}$$

where $k = 2\pi/\lambda$ is the wave number.

Representing now the spatial dependences of the fields $\mathcal{E}_1$ and $\mathcal{E}_2$ in the form $\mathcal{E}_1 = A_1 e^{-i\mathbf{k}_1\mathbf{r}}$ and $\mathcal{E}_2 = A_2 e^{-i\mathbf{k}_2\mathbf{r}}$, neglecting diffraction spreading, taking account of (1.4), and using a standard method [32], one can easily obtain from the wave equation the expression

$$\sum_{j=1}^{2}\left(\mathbf{k}_j \frac{\partial A_j}{\partial r} + \frac{1}{c} k \frac{\partial A_j}{\partial t}\right) e^{-i\mathbf{k}_j\mathbf{r}} + \text{c. c.} = -\frac{1}{2} k(1 - i\delta) g \sum_{j=1}^{2} A_j e^{-i\mathbf{k}_j\mathbf{r}} + \text{c. c.} \tag{1.5}$$

(summation is taken over the fields of two counter-propagating waves).

Introducing the saturation intensity of the two-level medium (in quanta), $I_s = 1/2\,\sigma T_1$, and setting $I_s = \mathcal{E}_s \mathcal{E}_s^*$, later on we change to the normalized field amplitudes $e_j = \mathcal{E}_j / \mathcal{E}_s$ and $a_j = A_j / \mathcal{E}_s$. Since $I / I_s = ee^*$, equation (1.2) gives

$$\frac{\partial g}{\partial t} + \frac{1}{T_1}(g - \alpha_0) = -\frac{1}{T_1} gee^*, \tag{1.6}$$

where $e = \sum_{j=1}^{2} e_j$ is the total normalized electromagnetic field in the medium.

Let us choose the $z$ axis to be the direction of propagation of the wave $E_1$. Since $\mathbf{k}_1 + \mathbf{k}_2 = 0$ for counter-propagating waves, then

$$ee^* = a_1 a_1^* + a_2 a_2^* + a_1 a_2^* e^{-i2k_1 z} + a_1^* a_2 e^{i2k_1 z}. \tag{1.7}$$

In view of (1.6) and (1.7), we seek the expression for $g(z, t)$ in the form [33]

$$g(z,\ t) = g_0(t) + \sum_{m=1}^{\infty} [g_m(t)\, e^{-i2mk_1 z} + g_m^*(t)\, e^{i2mk_1 z}]. \tag{1.8}$$

Substituting (1.8) into (1.5) and collecting the items with identical rapidly varying factors $e^{-ik_1 z}$ and $e^{ik_1 z}$ in the left and the right sides of the latter equation, we obtain the equations for slowly varying amplitudes

$$\frac{\partial a_1}{\partial z} + \frac{1}{c}\frac{\partial a_1}{\partial t} = -\frac{1}{2}(1 - i\delta)(g_0 a_1 + g_1 a_2), \tag{1.9}$$
$$-\frac{\partial a_2}{\partial z} + \frac{1}{c}\frac{\partial a_2}{\partial t} = -\frac{1}{2}(1 - i\delta)(g_1^* a_1 + g_0 a_2).$$

In the same way, using (1.6)-(1.8), we find the infinite chain of bound equations for the coefficients $g_m(t)$:

$$\frac{\partial g_0}{\partial t}+\frac{1}{T_1}(g_0-\alpha_0)=-\frac{1}{T_1}[(a_1a_1^*+a_2a_2^*)g_0+a_1^*a_2g_1+a_1a_2^*g_1^*],$$
$$\frac{\partial g_k}{\partial t}+\frac{1}{T_1}g_k=-\frac{1}{T_1}[(a_1a_1^*+a_2a_2^*)g_k+a_1^*a_2g_{k+1}+a_1a_2^*g_{k-1}], \quad (1.10)$$
$$k=1,\ 2,\ 3,\dots$$

We shall dwell only on the case of stationary interaction without carrying out a general analysis of the equations obtained. Then, setting $\partial/\partial t = 0$, we reduce (1.9) to the form

$$da_1/dz=-{}^1\!/_2\,(1-i\delta)\,(g_0a_1+g_1a_2), \quad (1.11)$$
$$-da_2/dz=-{}^1\!/_2\,(1-i\delta)\,(g_1^*a_1+g_0a_2).$$

Then set of equations (1.10) gives

$$(1+I_1+I_2)\,g_0+pg_1+p^*g_1^*=\alpha_0, \quad (1.12)$$
$$(1+I_1+I_2)\,g_k+pg_{k+1}+p^*g_{k-1}=0.$$
$$k=1,\ 2,\ 3,\dots,$$

where $p = a_1^*a_2$ and $I_k = a_ka_k^*$ are the normalized intensities of interacting waves.

Setting $g_k = C\lambda^k$, we obtain from the second equation of (1.12)

$$\lambda^2+\frac{1}{p}(1+I_1+I_2)\lambda+p^*\frac{1}{p}=0.$$

The roots $\lambda_1$ and $\lambda_2$ of this equation satisfy the relation $\lambda_1\lambda_2^* = 1$ and, moreover,

$$\lambda_{1,2}=\frac{1}{2p}(1+I_1+I_2)\left[\sqrt{1-\frac{4I_1I_2}{(1+I_1+I_2)^2}}\mp 1\right].$$

Choosing $\lambda_1$ to satisfy the inequality $|\lambda_1| < 1$, we obviously obtain $|\lambda_2| > 1$. And since the general solution of homogeneous set of equations (1.12) can be represented in the form $g_k = C_1\lambda_1^k + C_2\lambda_2^k$ ($k$ = 0, 1, 2, ...), where $C_1$ and $C_k$ are some constants, it is necessary to set $C_2$ = 0 for the limited solution $g_k$. As a result, the first equation of (1.12) gives

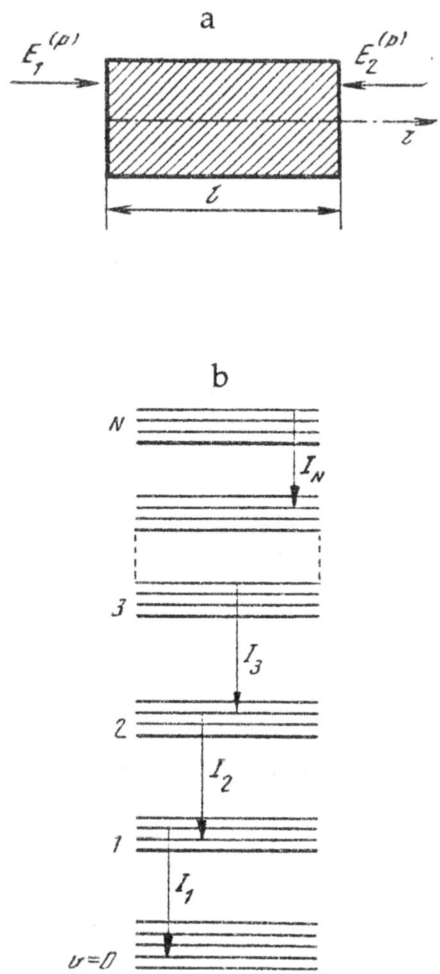


Fig. 1. Simplified configuration of two-wave interaction of counter--propagating waves (a) and general diagram of cascade transitions in a multilevel resonant medium (b).

$$C_1 = \frac{\alpha_0}{(1 + I_1 + I_2) + p\lambda_1^* + p^*\lambda_1}.$$

Thus

$$g_0 = \frac{\alpha_0}{\Delta}, \quad g_1 = \frac{\alpha_0}{2a_1^* a_2}\left(1 - \frac{1 + I_1 + I_2}{\Delta}\right), \tag{1.13}$$

where

$$\Delta = [(1 + I_1 + I_2)^2 - 4I_1 I_2]^{1/2}.$$

Substituting relations (1.13) into (1.11), we obtain the final equations describing interference interaction of counter-propagating waves in a two-level absorbing medium

$$\frac{da_1}{dz} = -\frac{(1 - i\delta)\,\alpha_0 a_1}{2\Delta}\left\{1 + \frac{1}{2I_1}[\Delta - (1 + I_1 + I_2)]\right\}, \tag{1.14}$$
$$-\frac{da_2}{dz} = -\frac{(1 - i\delta)\,\alpha_0 a_2}{2\Delta}\left\{1 + \frac{1}{2I_2}[\Delta - (1 + I_1 + I_2)]\right\}.$$

They are well-known, therefore we do not discuss them and pass on straight to the analysis of a more general case of interference interaction of mutifrequency radiation.

## 1.2 Equations of two-wave interaction of multifrequency radiation with widely separated frequencies in multilevel resonant media

Let us assume that multifrequency radiation propagating in opposite directions along the $Z$ axis is incident from two sides on an amplifying (absorbing) multilevel medium. This radiation is supposed to be resonantly amplified (absorbed) on separate vibrational-rotational transitions of various vibrational bands (Fig. 1a, b). This case is realized, for example, in the cavities of gas lasers working on cascade transitions of vibrational bands of two-level molecules such as HF, CO, etc. Because of their appreciable anharmonicity, the radiation of such laser systems consists of a number of separate lines that do not overlap.

Supposing that the rotational relaxation is fast, later on we shall consider the Boltzmann distribution of molecules over the rotational sublevels. Then, taking account of the pumping processes, vibrational relaxation, radiative transitions, and spatial diffusion of excited molecules ( their spatial distribution appears to be inhomogeneous because of the interference of counter-propagating waves), the equations for population of separate vibrational levels, $n_k$, can be written in the form

$$\frac{\partial n_k}{\partial t} = -(1-\delta_{0,k})\left(\frac{n_k}{T_k}+g_k I_k\right)+ \\ +(1-\delta_{N,k})\left(\frac{n_{k+1}}{T_{k+1}}+g_{k+1}I_{k+1}\right)+\alpha_k W(t)+D\nabla^2 n_k, \qquad k=0,1,\ldots,N. \tag{1.15}$$

Here $n_k$ is the density of molecules on the $k$ th vibrational level; $t$ the time; $W$ the excitation rate produced by an external source (electric discharge, chemical pumping, etc.); $\alpha_k$ and $T_k$ the probability of excitation and the characteristic relaxation time of the $k$ th level, respectively; $g_k = x_k \sigma_k (n_k - \beta_{j_k} n_{k-1})$ the gain for the vibrational-rotational transition $k, j_k \to k-1, j_{k-1}$; $\sigma_k$ the cross section (frequency-dependent in the general case) for radiative transition; $x_k = (\theta_r/T)(2j_k+1)\exp\{-(\theta_r/T)j_k(j_k+1)$ the rotational Boltzmann factor ( $\theta_r$ the characteristic rotational temperature, $T$ the gas temperature); $I_k$ the total intensity of counter-propagating waves with the frequencies $\omega_k$ ; $\delta_{ik}$ the Kroneker symbol; $\beta_{j_k} = \exp\{-(\theta_r/T)[j_{k-1}(j_{k-1}+1)-j_k(j_k+1)]\}$, and $D$ the diffusion coefficient of excited molecules.

Since $\sum_{k=0}^{N} n_k = \int_0^t W dt = N_0$ ( $N_0$ is the total density of molecules; in particular, $N_0$ = const if the the number of molecules time $T_\varphi$ is usually small in comparison with the vibrational relaxation times $T_k$ ( $k$ = 1, 2, ..., $N$ ). That is why, if the pulse durations of interacting waves are not too short, one can use the quasistatic approximation for slowly varying in time complex amplitudes of polarization at separate frequencies, setting $\partial \mathcal{P}^{(p)}/\partial t \simeq 0$.

Since the gas lasers often work under the conditions of fast translational–translational molecular relaxation (in comparison with the rates of other characteristic processes), then the effects of "hole burning" in the gain (absorption) profiles are usually not very important even in the case of inhomogeneous broadening of spectral lines. In this situation we may use the following expression for resonant polarization $\mathcal{P}^{(p)}$ on a separate radiative transition $p, j_p \rightarrow p-1, j_{p-1}$ [34]:

$$\mathcal{P}^{(p)} = \frac{1}{4\pi q^{(p)}} g_p \mathcal{E}^{(p)} \frac{\Phi(x_p, y_p)}{F(x_p, y_p)}, \tag{1.17}$$

where $q^{(p)}$ is the wave number for the field with the frequency $\omega_p$ ; $g_p$ the frequency dependent gain on the transition being considered; and $\Phi(x,y) = L(x,y) + iF(x,y)$ the complex Foigt function. Here

$$L(x, y) = \frac{1}{\pi} \int_{-\infty}^{\infty} \frac{(x-t)\, e^{-t^2}\, dt}{(x-t)^2 + y^2}, \quad F(x, y) = \frac{y}{\pi} \int_{-\infty}^{\infty} \frac{e^{-t^2}\, dt}{(x-t)^2 + y^2}.$$

In addition, the designations $x_p = (\omega_p - \Omega_p)\sqrt{\ln 2}/\Delta\omega_D$ and $y_p = \Delta\omega_L\sqrt{\ln 2}/\Delta\omega_D$ are used in (1.17). Here $\Omega_p$ is the transition frequency at the line center; $\Delta\omega_D$ and $\Delta\omega_L$ the half-widths of Doppler and Lorentz lines, respectively.

It should be noted that in the case of homogeneous broadening ( $\Delta\omega_D \ll \Delta\omega_L$ ), using the asymptotic expressions for the functions $L(x,y)$ and $F(x,y)$ for $y \rightarrow \infty$ and taking into account that per unit volume does not depend on pumping) and $n_k - \beta_{jk} n_{k-1}$ = $g_k / \varkappa_k \sigma_k$, then all $n_k$ in (1.15) can be expressed through $g_k$ and $N_0$. Then we can go from the set of equations for $n_k(t)$ to the set for $g_k(t)$.

Introducing the saturation intensities for separate radiative transitions in the standard way, $I_s^{(k)} = 1/(1+\beta_{jk})\varkappa_k \sigma_k T_k$, we normalize the intensities of all the waves setting $\tilde{I}_k = I_k / I_s^{(k)}$. As a result, we obtain on the basis of (1.15)

$$\frac{\partial g_k}{\partial t} + \sum_{p=1}^{N} f_p^{(k)} g_p + (1-\delta_{1,k})\, Q_{k-1}^{(k)} g_{k-1} \tilde{I}_{k-1} + \qquad (1.16)$$

$$+ Q_k^{(k)} g_k \tilde{I}_k + (1-\delta_{N,k})\, Q_{k+1}^{(k)} g_{k+1} \tilde{I}_{k+1} = R_k + D\nabla^2 g_k,$$

$$k = 1, 2, \ldots, N.$$

It is obvious that the concrete form of coefficients $f_p^{(k)}$, $Q_p^{(k)}$, and $R_k$ in (1.16) depends on the type of a pump source, the radiation spectrum, and the characteristic relaxation times ( see, for example, section 1.3).

Turning to the wave equations for the radiative field (1.3), we represent the total field in the form of superposition of all the fields incident on the medium. Then $E = \sum_{p=1}^{N} E^{(p)}$, where $E^{(p)} = E_1^{(p)} + E_2^{(p)}$ is the total electromagnetic field of two counter-propagating waves with the frequency $\omega_p$.

Using the temporal dependences in the form $E^{(p)} = \mathcal{E}^{(p)} e^{i\omega_p t}$ + c.c. for the fields of separate frequencies and, moreover, supposing that the spectral line profiles do not overlap and that the difference ( $\omega_p - \omega_k$ ) and summary ( $\omega_p + \omega_k$ ) frequencies are far from the intrinsic characteristic molecular frequencies, we seek the resonant polarization of multilevel system in the form

$$P = \sum_{p=1}^{N} P^{(p)}, \text{ где } P^{(p)} = \mathscr{P}^{(p)} e^{i\omega_p t} + \text{c. c.}$$

In molecular media, the characteristic polarization relaxation

$\Delta\omega_L = 1/T_\varphi$ , we obtain from (1.17) the expression that is analogous to (1.4):

$$\mathcal{P}^{(p)} = \frac{i + \delta_p}{4\pi q^{(p)}} g_p \mathcal{E}^{(p)}, \tag{1.18}$$

where $\delta_p = (\omega_p - \Omega_p) T_\varphi$ is the frequency detuning.

As for the case of exact resonance ($\omega_p = \Omega_p$), it follows from (1.17) that the expression for $\mathcal{P}^{(p)}$ has the same form for both homogeneous and inhomogeneous line broadening, however the cross section for radiative transition is, of course, determined with regard for the type of broadening.

Because of the linear superposition of polarizations with different frequencies, the equations for $E^{(p)}$ in (1.3) can be separated. Therefore, using the standard methods, one can easily obtain the shortened equations for slowly varying amplitudes of these fields. For this purpose we separate out the rapidly oscillating (in space) terms in $\mathcal{E}^{(p)}$ setting

$$\mathcal{E}^{(p)} = A_1^{(p)} e^{-i\mathbf{q}_1^{(p)}\mathbf{r}} + A_2^{(p)} e^{-i\mathbf{q}_2^{(p)}\mathbf{r}} \tag{1.19}$$

(the wave vectors of counter-propagating waves with equal frequencies satisfy the phase matching condition $\mathbf{q}_1^{(p)} + \mathbf{q}_2^{(p)} = 0$).

Restricting ourselves for simplicity only to the case of exact resonance ($\delta_p = 0$), we substitute (1.18) and (1.19) into the wave equation for the radiation field of corresponding frequency. As a result, neglecting the diffraction spreading of the fields in the media as before, we obtain the equations for the amplitudes $A_1^{(p)}$ and $A_2^{(p)}$ :

$$\sum_{l=1}^{2} \left( \mathbf{q}_l^{(p)} \frac{\partial A_l^{(p)}}{\partial \mathbf{r}} + \frac{1}{c} q^{(p)} \frac{\partial A_l^{(p)}}{\partial t} \right) e^{-i\mathbf{q}_l^{(p)}\mathbf{r}} = \\ = \frac{1}{2} q^{(p)} g_p \sum_{l=1}^{2} A_l^{(p)} e^{-i\mathbf{q}_l^{(p)}\mathbf{r}}, \qquad p = 1, 2, \ldots, N. \tag{1.20}$$

Dividing both parts of (1.20) by $(I_s^{(p)})^{1/2}$, it is always possible to normalize the wave amplitudes. Therefore, appropriately normalized $A_\ell^{(p)}$ will be considered below without any special notes. Using set of equations (1.16) and (1.20), it is easy to obtain the equations of interference interaction for the case of multifrequency radiation by the method analogous to the one described in section 1.1 [35].

Since $\tilde{I}_p = \mathcal{E}^{(p)}\mathcal{E}^{(p)*}$, we obtain from (1.19)

$$\tilde{I}_p = A_1^{(p)}A_1^{(p)*} + A_2^{(p)}A_2^{(p)*} + A_1^{(p)}A_2^{(p)*}e^{-i2q^{(p)}z} + A_2^{(p)*}A_2^{(p)}e^{i2q^{(p)}z}. \quad (1.21)$$

Taking account of (1.21) and (1.16), we seek the expressions for gain $g_k(z,t)$ in the form of the following expansions in spatial harmonics of the fields at all the frequencies:

$$g_k(z,t) = g_{0,k}^{(k)} + \sum_{p=1}^{N}\sum_{m=1}^{\infty}\left(g_{2m,k}^{(p)}e^{-i2mq^{(p)}z} + \text{c.c.}\right), \quad (1.22)$$
$$k = 1, 2, \ldots, N.$$

Substituting (1.22) into (1.20) and grouping the terms with identical rapidly varying factors of the form $\exp(\pm iq^{(p)}z)$ in the left and right sides of the latter equation, we obtain the following equations for slowly varying amplitudes of separate waves:

$$\frac{\partial A_1^{(p)}}{\partial z} + \frac{1}{c}\frac{\partial A_1^{(p)}}{\partial t} = \frac{1}{2}\left(g_{0,p}^{(p)}A_1^{(p)} + g_{2,p}^{(p)}A_2^{(p)}\right),$$
$$-\frac{\partial A_2^{(p)}}{\partial z} + \frac{1}{c}\frac{\partial A_2^{(p)}}{\partial t} = \frac{1}{2}\left(g_{2,p}^{(p)*}A_1^{(p)} + g_{0,p}^{(p)}A_2^{(p)}\right), \quad (1.23)$$
$$p = 1, 2, \ldots, N.$$

When deducing equation (1.23), it was assumed that the frequencies $\omega_p$ of separate lines were considerably different, so that $|\Delta_{k,p}| = |q^{(k)} - q^{(p)}| \gg |g_{0,k}^{(k)}|$ (the case of widely separated frequencies). In this case one may neglect the terms with rapidly oscillating factors of the form $e^{\pm i2\Delta_{k,p}z}$ in the right sides of Eqs. (1.23) because the variation of field amplitudes and coefficients $g_{m,p}^{(k)}$ in the medium is considerably slower.

In the same way we obtain the infinite chain of equations for coefficients $g_{2m,k}^{(p)}(t)$ from (1.16). So, the constant components $g_{0,k}^{(k)}$ in expression (1.22) are given by

$$\frac{\partial g_{0,k}^{(k)}}{\partial t}+\sum_{l=1}^{N} f_l^{(k)} g_{0,l}^{(l)}+(1-\delta_{1,k})\,Q_{k-1}^{(k)}\,[C_{0,0}^{(k-1)} g_{0,k-1}^{(k-1)}+C_{2,0}^{(k-1)*} g_{2,k-1}^{(k-1)}+ \\ +C_{2,0}^{(k-1)} g_{2,k-1}^{(k-1)*}]+Q_k^{(k)}\,[C_{0,0}^{(k)} g_{0,k}^{(k)}+C_{2,0}^{(k)*} g_{2,k}^{(k)}+C_{2,0}^{(k)} g_{2,k}^{(k)*}]+ \\ +(1-\delta_{N,k})\,Q_{k+1}^{(k)}\,[C_{0,0}^{(k+1)} g_{0,k+1}^{(k+1)}+C_{2,0}^{(k+1)*} g_{2,k+1}^{(k+1)}+C_{2,0}^{(k+1)} g_{2,k+1}^{(k+1)*}]=R_k, \tag{1.24}$$

When $m > 0$, we obtain the following expressions for coefficients $g_{2m,k}^{(p)}$ (here $k = 1, 2, \ldots, N$ ; $m = 1, 2, \ldots$):

$$\frac{\partial g_{2m,k}^{(p)}}{\partial t}+\frac{1}{\tau_{2m}^{(p)}}\, g_{2m,k}^{(p)}+\sum_{l=1}^{N} f_l^{(k)} g_{2m,l}^{(p)}+(1-\delta_{1,k})\,Q_{k-1}^{(k)} C_{0,0}^{(k-1)} g_{2m,k-1}^{(p)}+ \\ +Q_k^{(k)} C_{0,0}^{(k)} g_{2m,k}^{(p)}+(1-\delta_{N,k})\,Q_{k+1}^{(k)} C_{0,0}^{(k+1)} g_{2m,k+1}^{(p)}=0, \tag{1.25}$$

where $p = 1, 2, \ldots, N$ and $p \neq k-1, k, k+1$;

$$\frac{\partial g_{2m,k}^{(k-1)}}{\partial t}+\frac{1}{\tau_{2m}^{(k-1)}}\, g_{2m,k}^{(k-1)}+\sum_{l=1}^{N} f_l^{(k)} g_{2m,l}^{(k-1)}+(1-\delta_{1,k})\,Q_{k-1}^{(k)}\times \\ \times[C_{0,0}^{(k-1)} g_{2m,k-1}^{(k-1)}+C_{2,0}^{(k-1)*} g_{2m+2,k-1}^{(k-1)}+C_{2,0}^{(k-1)} g_{2m-2,k-1}^{(k-1)}]+ \\ +Q_k^{(k)} C_{0,0}^{(k)} g_{2m,k}^{(k-1)}+(1-\delta_{N,k})\,Q_{k+1}^{(k)} C_{0,0}^{(k+1)} g_{2m,k+1}^{(k-1)}=0, \tag{1.26}$$

$$\frac{\partial g_{2m,k}^{(k)}}{\partial t}+\frac{1}{\tau_{2m}^{(k)}}\, g_{2m,k}^{(k)}+\sum_{l=1}^{N} f_l^{(k)} g_{2m,l}^{(k)}+(1-\delta_{1,k})\,C_{0,0}^{(k-1)} g_{2m,k-1}^{(k)}+ \\ +Q_k^{(k)}\,[C_{0,0}^{(k)} g_{2m,k}^{(k)}+C_{2,0}^{(k)*} g_{2m+2,k}^{(k)}+C_{2,0}^{(k)} g_{2m-2,k}^{(k)}]+ \\ +(1-\delta_{N,k})\,Q_{k+1}^{(k)} C_{0,0}^{(k+1)} g_{2m,k+1}^{(k)}=0, \tag{1.27}$$

$$\frac{\partial g_{2m,k}^{(k+1)}}{\partial t}+\frac{1}{\tau_{2m}^{(k+1)}}\, g_{2m,k}^{(k+1)}+\sum_{l=1}^{N} f_l^{(k)} g_{2m,l}^{(k+1)}+(1-\delta_{1,k})\,Q_{k-1}^{(k)} C_{0,0}^{(k-1)} g_{2m,k-1}^{(k+1)}+ \\ +Q_k^{(k)} C_{0,0}^{(k)} g_{2m,k}^{(k+1)}+(1-\delta_{N,k})\,Q_{k+1}^{(k)}\,[C_{0,0}^{(k+1)} g_{2m,k+1}^{(k+1)}+ \\ +C_{2,0}^{(k+1)*} g_{2m+2,k+1}^{(k+1)}+C_{2,0}^{(k+1)} g_{2m-2,k+1}^{(k+1)}]=0. \tag{1.28}$$

Eqs. (1.24)-(1.28) contain the following designations

$$C_{0,0}^{(k)} = A_1^{(k)} A_1^{(k)*} + A_2^{(k)} A_2^{(k)*},$$
$$C_{2,0}^{(k)} = A_1^{(k)} A_2^{(k)*}, \qquad C_{2,0}^{(k)*} = A_1^{(k)*} A_2^{(k)}. \tag{1.29}$$

The quantities $\tau_{2m}^{(p)} = 1/4\,(q^{(p)})^2 D m^2$ characterize the times of diffusive spreading of amplitude gratings with the spatial period $\lambda_p/2m$ that are formed as a result of interference of counter-propagating waves having the frequency $\omega_p$. It was additionally assumed in (1.24)-(1.28) that $g_{2m,k}^{(p)} = 0$ when $k, p \leq 0$ and $k, p > N$.

So, set of equations for field amplitudes (1.23) in combination with the infinite (in $m$) chain of equations (1.24)-(1.28) describe the nonstationary two-wave interaction of multifrequency laser radiation in multilevel media. It is clear that in the general case such a system can be studied only by numerical methods. In concrete calculations, the infinite chain of equations for coefficients $g_{2m,k}^{(p)}$ must be shortened to a certain finite number of components in expansion (1.22).

It should be noted that if the diffusive spreading of the gratings is neglected (this is possible, for example, in the strong fields for sufficiently short pulse durations of the interacting waves ( $t_p \lesssim 1/Dq^2$ )), then in principle a set of closed equations for the field amplitudes can be obtained in the quasistationary case ( $\partial/\partial t = 0$) from the equations being analyzed.

Indeed, setting $p = 1$, let us write out the equations for $g_{2m,k}^{(1)}$ for $k = 3, 4, \ldots, N$ (we suppose that $p \neq k$, $k - 1$) on the basis of (1.25):

$$\sum_{l=1}^{N} f_l^{(3)} g_{2m,l}^{(1)} + Q_2^{(3)} C_{0,0}^{(2)} g_{2m,2}^{(1)} + Q_3^{(3)} C_{0,0}^{(3)} g_{2m,3}^{(1)} + Q_4^{(3)} C_{0,0}^{(4)} g_{2m,4}^{(1)} = 0,$$
$$\sum_{l=1}^{N} f_l^{(4)} g_{2m,l}^{(1)} + Q_3^{(4)} C_{0,0}^{(3)} g_{2m,3}^{(1)} + Q_4^{(4)} C_{0,0}^{(4)} g_{2m,4}^{(1)} + Q_5^{(4)} C_{0,0}^{(5)} g_{2m,5}^{(1)} = 0, \tag{1.30}$$
$$\cdots\cdots\cdots\cdots\cdots\cdots\cdots\cdots$$
$$\sum_{l=1}^{N} f_l^{(N)} g_{2m,l}^{(1)} + Q_{N-1}^{(N)} C_{0,0}^{(N-1)} g_{2m,N-1}^{(1)} + Q_N^{(N)} C_{0,0}^{(N)} g_{2m,N}^{(1)} = 0.$$

In order to close the system with respect to coefficients $g^{(1)}_{2m,r}$ ( $r$ = 1, 2, ..., $N$ ), let us also use Eq. (1.26) for $k$ = 2 and Eq. (1.27) for $k$ = 1

$$\sum_{l=1}^{N} f_l^{(1)} g^{(1)}_{2m,\,l} + Q_1^{(1)}\,[C^{(1)}_{0,\,0} g^{(1)}_{2m,\,1} + C^{(1)*}_{2,\,0} g^{(1)}_{2m+2,\,1} + C^{(1)}_{2,\,0} g^{(1)}_{2m-2,\,1}] + $$
$$+ Q_2^{(1)} C^{(2)}_{0,0} g^{(1)}_{2m,\,2} = 0,$$
$$\sum_{l=1}^{N} f_l^{(2)} g^{(1)}_{2m,\,l} + Q_1^{(2)}\,[C^{(1)}_{0,\,0} g^{(1)}_{2m,\,1} + C^{(1)*}_{2,\,0} g^{(1)}_{2m+2,\,1} + C^{(1)}_{2,\,0} g^{(1)}_{2m-2,\,1}] + \tag{1.31}$$
$$+ Q_2^{(2)} C^{(2)}_{0,\,0} g^{(1)}_{2m,\,2} + Q_3^{(2)} C^{(3)}_{0,\,0} g^{(1)}_{2m,\,3} = 0.$$

Combining either of Eqs. (1.31) with set (1.30), one can express $N$ - 1 coefficients $g^{(1)}_{2m,\,r}$ ( $r \neq 1$) of the resulting set of ( $N$ -1) equations in $N$ unknowns through $g^{(1)}_{2m,1}$ in a linear form and then substitute the expressions obtained into the last equation remaining in (1.31). As a result, we obtain the following equation for $g^{(1)}_{2m,1}$

$$\xi g^{(1)}_{2m,\,1} + \eta\,(C^{(1)*}_{2,\,0} g^{(1)}_{2m+2,\,1} + C^{(1)}_{2,\,0} g^{(1)}_{2m-2,\,1}) = 0, \tag{1.32}$$

where $m$ = 1, 2, ..., while $\xi$ and $\eta$ are certain real functions of $C^{(k)}_{0,0}$ (interacting wave intensities).

The chain of equations (1.32) can be easily solved. Let us set $y_m = g^{(1)}_{2m,1}$ and seek the solution for $y_m$ ( $m$ = 0, 1, ...) in the form $y_m = S\lambda^m$, where $S$ is a constant. Then (1.32) gives the quadratic equation for:

$$\lambda^2 + \frac{\xi}{\eta C^{(1)*}_{2,\,0}}\,\lambda + \frac{C^{(1)}_{2,\,0}}{C^{(1)*}_{2,\,0}} = 0, \tag{1.33}$$

with the roots $\lambda_1$ and $\lambda_2$ satisfying the relation $\lambda_1\lambda_2^* = 1$. Therefore, if $|\lambda_1| < 1$, then $|\lambda_2| > 1$ and

$$\lambda_{1,\,2} = -\frac{\xi}{2\eta C^{(1)*}_{2,\,0}}\,(1 \mp \Delta),$$

where $\Delta = (1 - 4\eta^2 C_{2,0}^{(1)} C_{2,0}^{(1)*}/\xi^2)^{1/2}$. Being interested only in the limited solution of set (1.32), we reject the growing solution with $\lambda = \lambda_2$. Then $g_{2m,1}^{(1)} = \delta_1 \lambda_1^m$ ( $m$ = 0, 1, ...).

It is clear that one can find the expressions for all the other coefficients $g_{2m,k}^{(k)}$ ( $k$ = 2, 3, ..., $N$ ) in the same way. Then $g_{2m,k}^{(k)} = \delta_k \lambda_k^m$, where $\delta_k$ are the constants that should be determined from the corresponding boundary conditions. We note that the latter are obtained directly from (1.24) when $\partial/\partial t = 0$. Since all $R_k$ ( $k$ = 1, 2, ..., $N$ ) are the real quantities, then all $\delta_k$ are the real quantities in this case too. Solving the set of equations for $\delta_1$, $\delta_2$, ..., $\delta_N$, one can find the concrete values of all coefficients $g_{2m,k}^{(k)}$ expressed in a definite form through the intensities of interacting waves and the coefficients of set (1.16).

So, it directly follows from the above-stated analysis that the approach applied gives in the stationary (or quasistationary) case the exact relation between coefficients of two-wave interaction and the wave intensities (amplitudes) for any number of spectral components of the radiation fied. In the next section we shall illustrate the deduction of such equations and some characteristic consequences obtained with due regard for interference effects by the concrete examples of two- and three-frequency two-wave interactions.

## 1.3. Some illustrations

### 1.3.1. Two-frequency radiation

Restricting ourselves for simplicity to the case of a three-level molecular medium, we set $N$ = 2 in (1.15). Let us consider the concrete scheme of pumping the vibrational levels in an electric discharge. Then the excitation rate is given by the relation $W(t) = n_0(t)/T_*$, where $n_0$ is the population density of the lower vibrational level, $T_* = 1/k_e n_e$ the characteristic excitation time ($k_e$ the rate constant for excitation of molecules by electron impact), $n_e$ the electron density in the discharge. In this case the probabilities of excitation of separate levels are given by $\alpha_0 = -1$ and $\alpha_1 + \alpha_2 = 1$. Therefore, using the dimensionless time $\tau = t/T_1$ ($T_1$ is the population relaxation time for the first vibrational level) in (1.15) for sake of convenience, we obtain the equations for gain $g_1$ and $g_2$ at the frequencies $\omega_1$ and $\omega_2$ respectively (compare with (1.16))

$$\frac{\partial g_1}{\partial \tau} + f_1^{(1)} g_1 + f_2^{(1)} g_2 + Q_1^{(1)} g_1 \tilde{I}^{(1)} + Q_2^{(1)} g_2 \tilde{I}^{(2)} = R_1 + DT_1 \nabla^2 g_1,$$
$$\frac{\partial g_2}{\partial \tau} + f_1^{(2)} g_1 + f_2^{(2)} g_2 + Q_1^{(2)} g_1 \tilde{I}^{(1)} + Q_2^{(2)} g_2 \tilde{I}^{(2)} = R_2 + DT_1 \nabla^2 g_2. \quad (1.34)$$

One can show that the coefficients of set (1.34) have the form

$$\begin{aligned}
R_1 &= x_1 \sigma_1 N_0 [\tau_* (\alpha_1 + \beta_{j_1}) + \tau_1 \beta_{j_1} \beta_{j_2} - \beta_{j_1} (1 + \beta_{j_1})]/s_0,\\
R_2 &= x_2 \sigma_2 N_0 [\tau_* (\alpha_2 - \beta_{j_2} \alpha_1) - \tau_1 \beta_{j_1} \beta_{j_2} (1 + \beta_{j_2}) + \beta_{j_1} \beta_{j_2}]/s_0,\\
f_1^{(1)} &= [\tau_* (1 + \beta_{j_2})(\alpha_1 + \beta_{j_1}) - \tau_1 \beta_{j_2} + 1 + \beta_{j_1}]/s_0,\\
f_2^{(1)} &= [\tau_* (\alpha_1 + \beta_{j_1}) - \tau_1 (1 + \beta_{j_1}) - \beta_{j_1} (1 + \beta_{j_1})]/\sigma s_0,
\end{aligned} \quad (1.35)$$

$$\begin{aligned}
f_1^{(2)} &= \sigma [\tau_* (1 + \beta_{j_2})(\alpha_2 - \beta_{j_1} \alpha_1) + \tau_1 \beta_{j_2} (1 + \beta_{j_2}) - \beta_{j_2}]/s_0,\\
f_2^{(2)} &= [\tau_* (\alpha_2 - \beta_{j_2} \alpha_1) + \tau_1 (1 + \beta_{j_1})(1 + \beta_{j_2}) + \beta_{j_1} \beta_{j_2}]/s_0,\\
Q_1^{(1)} &= 1, \qquad Q_2^{(1)} = -\tau_1/\sigma (1 + \beta_{j_2}),\\
Q_1^{(2)} &= -\sigma \beta_{j_2}/(1 + \beta_{j_1}), \qquad Q_2^{(2)} = \tau_1,
\end{aligned}$$

where $\tau_* = T_1/T_*$, $s_0 = 1 + \beta_{j_1} + \beta_{j_1}\beta_{j_2}$, $\tau_1 = T_1/T_2$ and $\sigma = x_2\sigma_2/x_1\sigma_1$.

It should be noted that a passive (absorbing) medium ($W = 0$) is also described by Eqs. (1.34) if we set $\tau_* = 0$ in (1.35). Besides, the lower vibrational molecular levels, as a rule, are described satisfactorily by the model of a harmonic oscillator [36]. Therefore, one can take $\tau_1 = \sigma_2/\sigma_1 = 2$ as a good approximation. In particular, if the rotational molecular structure is neglected ($\beta_{j_1} = \beta_{j_2} = 1$,

$\mathcal{X}_1 = \mathcal{X}_2 = 1$), then it follows from (1.34) for a three-level absorbing medium

$$\begin{aligned} &\partial g_1/\partial\tau - g_2 + g_1\tilde{I}^{(1)} - {}^1\!/_2 g_2\tilde{I}^{(2)} = DT_1\nabla^2 g_1, \\ &\partial g_2/\partial\tau + 2g_1 + 3g_2 - g_1\tilde{I}^{(1)} + 2g_2\tilde{I}^{(2)} = -2\sigma_1 N_0 + DT_1\nabla^2 g_2. \end{aligned} \tag{1.36}$$

In the stationary mode ($\partial/\partial t = 0$) and in the case when the spatial diffusion of the particles can be neglected (high pressures, short vibrational relaxation times), we find from (1.36)

$$\begin{aligned} g_1 &= -\frac{2\sigma_1 N_0(2+\tilde{I}^{(2)})}{3\tilde{I}^{(1)}\tilde{I}^{(2)} + 4\tilde{I}^{(1)} + 2\tilde{I}^{(2)} + 4}, \\ g_2 &= -\frac{4\sigma_1 N_0\tilde{I}^{(1)}}{3\tilde{I}^{(1)}\tilde{I}^{(2)} + 4\tilde{I}^{(1)} + 2\tilde{I}^{(2)} + 4}. \end{aligned} \tag{1.37}$$

We shall return to the expressions obtained somewhat later and now let us examine the equations describing the stationary two-wave (interference) interaction of two-frequency radiation in absorbing and amplifying media [37]. According to (1.23), the equations for slowly varying complex amplitudes of counter-propagating waves $A_1^{(p)}$ and $A_2^{(p)}$ with the frequency $\omega_p$ ($p$ = 1, 2) can be represented in the form

$$\begin{aligned} dA_1^{(1)}/dz &= {}^1\!/_2\,(g_{0,1}^{(1)}A_1^{(1)} + g_{2,1}^{(1)}A_2^{(1)}), \\ -dA_2^{(1)}/dz &= {}^1\!/_2\,(g_{2,1}^{(1)*}A_1^{(1)} + g_{0,1}^{(1)}A_2^{(1)}) \end{aligned} \tag{1.38}$$

and

$$\begin{aligned} dA_1^{(2)}/dz &= {}^1\!/_2\,(g_{0,2}^{(2)}A_1^{(2)} + g_{2,2}^{(2)}A_2^{(2)}), \\ -dA_2^{(2)}/dz &= {}^1\!/_2\,(g_{2,2}^{(2)*}A_1^{(2)} + g_{0,2}^{(2)}A_2^{(2)}) \end{aligned} \tag{1.39}$$

We are reminded that the quantities $g_{0,1}^{(1)}$ and $g_{0,2}^{(2)}$ correspond to the constant components of gain (absorption) coefficients at the frequencies $\omega_1$ and $\omega_2$, respectively, in the expansion (1.22), while $g_{2m,k}^{(p)}$ characterize the amplitudes of spatial harmonics (caused by interference of counter-propagating waves having the frequency $\omega_p$ ) of gain $g_k$ on the transitions of 1 ⟶ 0 and 2 ⟶ 1 bands ( $k$ = 1, 2). Their periods are $\Lambda_m = \pi / m q^{(p)}$.

The infinite chain of Eqs. (1.25)–(1.28) for the coefficients of two-wave interaction gives

$$\begin{aligned}(f_1^{(k)} + Q_1^{(k)} C_{0,0}^{(1)} g_{2m,1}^{(1)} + Q_1^{(k)} (C_{2,0}^{(1)} g_{2m-2,1}^{(1)} + \\ + C_{2,0}^{(1)*} g_{2m+2,1}^{(1)}) = -(f_2^{(k)} + Q_2^{(k)} C_{0,0}^{(2)})\, g_{2m,2}^{(1)}\end{aligned} \tag{1.40}$$

and

$$\begin{aligned}-(f_1^{(k)} + Q_1^{(k)} C_{0,0}^{(1)})\, g_{2m,1}^{(2)} = (f_2^{(k)} + Q_2^{(k)} C_{0,0}^{(2)})\, g_{2m,2}^{(2)} + \\ + Q_2^{(k)} (C_{0,0}^{(2)} g_{2m-2,2}^{(2)} + C_{2,0}^{(2)*} g_{2m+2,2}^{(2)}),\end{aligned} \tag{1.41}$$

where $k$ = 1, 2 and $m$ = 1, 2, 3, ...

The relation between coefficients $g_{2m,1}^{(1)}$ and $g_{2m,2}^{(2)}$ is given by the boundary conditions that follow directly from (1.24):

$$\begin{aligned}&[(f_1^{(k)} + Q_1^{(k)} C_{\theta,0}^{(1)})\, g_{0,1}^{(1)} + Q_1^{(k)} (C_{2,0}^{(1)} g_{2,1}^{(1)*} + C_{2,0}^{(1)*} g_{2,1}^{(1)})] + \\ &+ [(f_2^{(k)} + Q_2^{(k)} C_{0,0}^{(2)})\, g_{0,2}^{(2)} + Q_2^{(k)} (C_{2,0}^{(2)} g_{2,2}^{(2)*} + C_{2,0}^{(2)*} g_{2,2}^{(2)})] = R_k, \\ &k = 1, 2.\end{aligned} \tag{1.42}$$

Taking account of (1.42), one can easily solve set of equations (1.40), (1.41). Indeed, at first let us address to Eqs. (1.40). Setting $k$ = 1 and 2 in succession, we eleminate the variables $g_{2m,2}^{(1)}$ from (1.40) and then for short introduce the designations

$$\begin{aligned}&\varkappa_1 = (f_2^{(1)} + Q_2^{(1)} C_{0,0}^{(2)}) / (f_2^{(2)} + Q_2^{(2)} C_{0,0}^{(2)}), \\ &\xi_1 = f_1^{(1)} - \varkappa_1 f_1^{(2)}, \quad \eta_1 = Q_1^{(1)} - \varkappa_1 Q_1^{(2)}, \quad y_m^{(1)} = g_{2m,1}^{(1)}.\end{aligned}$$

As a result, we get an expression for $y_m^{(1)}$

$$\begin{aligned}&(\xi_1 + \eta_1 C_{0,0}^{(1)})\, y_m^{(1)} + \eta_1 C_{2,0}^{(1)*} y_{m+1}^{(1)} + \eta_1 C_{2,0}^{(1)} y_{m-1}^{(1)} = 0, \\ &m = 1, 2, \ldots\end{aligned} \tag{1.43}$$

We seek the solution of (1.43) in the form $y_m^{(1)} = S \lambda^{(m)}$, where $S$ is some constant. Then a quadratic equation for $\lambda$ is obtained having the roots

$$\lambda_{1,2} = -\frac{\xi_1 + \eta_1 C_{0,0}^{(1)}}{2\eta_1 C_{2,0}^{(1)*}}(1 \mp \Delta_1), \tag{1.44}$$

where

$$\Delta_1 = [1 - 4\eta_1^2 C_{2,0}^{(1)} C_{2,0}^{(1)*}/(\xi_1 + \eta_1 C_{0,0}^{(1)})^2]^{1/2}.$$

In the same way introducing the designations

$$\varkappa_2 = (f_1^{(2)} + Q_1^{(2)} C_{0,0}^{(1)})/(f_1^{(1)} + Q_1^{(1)} C_{0,0}^{(1)}),$$
$$\xi_2 = f_2^{(2)} - \varkappa_2 f_2^{(1)}, \quad \eta_2 = Q_2^{(2)} - \varkappa_2 Q_2^{(1)}, \quad y_m^{(2)} = g_{2m,2}^{(2)},$$

we obtain from (1.41) the set of equations for $y_m^{(2)}$:

$$(\xi_2 + \eta_2 C_{0,0}^{(2)})\, y_m^{(2)} + \eta_2 C_{2,0}^{(2)*} y_{m+1}^{(2)} + \eta_2 C_{2,0}^{(2)} y_{m-1}^{(2)} = 0, \quad m = 1, 2, 3, \dots \tag{1.45}$$

Setting $y_m^{(2)} = S\Lambda^m$, where $S$ is a constant, we find from (1.45) two expressions for $\Lambda$:

$$\Lambda_{1,2} = -\frac{\xi_2 + \eta_2 C_{0,0}^{(2)}}{2\eta_2 C_{2,0}^{(2)*}}(1 \mp \Delta_2), \tag{1.46}$$

where $$\Delta_2 = [1 - 4\eta_2^2 C_{2,0}^{(2)} C_{2,0}^{(2)*}/(\xi_2 + \eta_2 C_{0,0}^{(2)})^2]^{1/2}.$$

It follows from (1.43) and (1.45) that $\lambda_1\lambda_2^* = \Lambda_1\Lambda_2^* = 1$. Therefore, we always have $|\lambda_2| > 1$ and $|\Lambda_2| > 1$ when $|\lambda_1| < 1$ and $|\Lambda_1| < 1$ (here the subscript "1" corresponds to the upper sign in (1.44) and (1.46)).

The general solutions of sets of homogeneous equations (1.43) and (1.45) have the form $y_m^{(1)} = S_1^{(1)}\lambda_1^m + S_2^{(1)}\lambda_2^m$ and $y_m^{(2)} = S_1^{(2)}\Lambda_1^m + S_2^{(2)}\Lambda_2^m$, where $S_k^{(p)}$ are some constants. Omitting the indefinitely growing solutions with $|\lambda_2| > 1$ and $|\Lambda_2| > 1$, we finally obtain

$$g^{(1)}_{2m,\,1} = S_1\lambda_1^m, \quad g^{(2)}_{2m,\,2} = S_2\Lambda_1^m, \qquad m = 0,\ 1,\ \ldots \tag{1.47}$$

In order to determine $S_1$ and $S_2$, we address to boundary conditions (1.42). It follows from the latter that $S_1$ and $S_2$ must be real. Therefore it is convenient to set

$$\begin{aligned}
d_1^{(1)} &= f_1^{(1)} + Q_1^{(1)}\left(C^{(1)}_{0,\,0} + C^{(1)}_{2,\,0}\lambda_1^* + C^{(1)*}_{2,\,0}\lambda_1\right),\\
d_2^{(1)} &= f_2^{(1)} + Q_2^{(1)}\left(C^{(2)}_{0,\,0} + C^{(2)}_{2,\,0}\Lambda_1^* + C^{(2)*}_{2,\,0}\Lambda_1\right),\\
d_1^{(2)} &= f_1^{(2)} + Q_1^{(2)}\left(C^{(1)}_{0,\,0} + C^{(1)}_{2,\,0}\lambda_1^* + C^{(1)*}_{2,\,0}\lambda_1\right),\\
d_2^{(2)} &= f_2^{(2)} + Q_2^{(2)}\left(C^{(2)}_{0,\,0} + C^{(2)}_{2,\,0}\Lambda_1^* + C^{(2)*}_{2,\,0}\Lambda_1\right).
\end{aligned}$$

As a result, we find from (1.42)

$$\begin{aligned}
S_1 &= \left(d_2^{(2)}R_1 - d_2^{(1)}R_2\right)/d,\\
S_2 &= \left(d_1^{(1)}R_2 - d_1^{(2)}R_1\right)/d,
\end{aligned} \tag{1.48}$$

where $d = d_1^{(1)}d_2^{(2)} - d_1^{(2)}d_2^{(1)}$.

Returning to (1.38) and (1.39) and taking account of (1.47), (1.44), (1.46), and (1.48), we find the final form of equations for two-frequency interference interaction:

$$\begin{aligned}
\frac{dA_1^{(1)}}{dz} &= \frac{1}{2}S_1A_1^{(1)}\left[1 + \frac{\xi_1 + \eta_1 C^{(1)}_{0,\,0}}{2\eta_1 A_1^{(1)}A_1^{(1)*}}(\Delta_1 - 1)\right],\\
-\frac{dA_2^{(1)}}{dz} &= \frac{1}{2}S_1A_2^{(1)}\left[1 + \frac{\xi_1 + \eta_1 C^{(1)}_{0,\,0}}{2\eta_1 A_2^{(1)}A_2^{(1)*}}(\Delta_1 - 1)\right]
\end{aligned} \tag{1.49}$$

and

$$\begin{aligned}
\frac{dA_1^{(2)}}{dz} &= \frac{1}{2}S_2A_1^{(2)}\left[1 + \frac{\xi_2 + \eta_2 C^{(2)}_{0,\,0}}{2\eta_2 A_1^{(2)}A_1^{(2)*}}(\Delta_2 - 1)\right],\\
-\frac{dA_2^{(2)}}{dz} &= \frac{1}{2}S_2A_2^{(2)}\left[1 + \frac{\xi_2 + \eta_2 C^{(2)}_{0,\,0}}{2\eta_2 A_2^{(2)}A_2^{(2)*}}(\Delta_2 - 1)\right].
\end{aligned} \tag{1.50}$$

Now let us return to the case of absorbing medium for which relations (1.37) are valid. We find from (1.37) that $g_1 \to -\sigma_1 N_0/(1+\tilde{I}'$

when $\tilde{I}^{(2)} \to 0$. It is obvious that this situation is equivalent to the case of a two-pass medium in the presence of single-frequency radiation. Therefore in the limit of zero radiation intensity at the frequency $\omega_2$ ( $A_1^{(2)}, A_2^{(2)} \to 0$), Eqs.(1.49) must change to Eqs.(1.14) describing two-wave interaction of single-frequency radiation. One can show that such a transition really takes place.

Unfortunately, it is impossible to obtain the analytical solution of set (1.49), (1.50). However, the role of interference of counter-propagating waves in the amplification (absorption) of incident radiation can be evaluated in the following way. Using relations (1.47), (1.48), (1.44), and (1.46), one can investigate, in principle, the dependence of coefficients $g_{0,k}^{(k)}$ and $g_{2,k}^{(k)}$ ($k$ = 1, 2) on the intensities of interacting waves. Figure 2 depicts such a dependence calculated for the three-level absorbing medium described by relations (1.37). In these calculations, equal intensities of all the waves ( $A_1^{(1)}A_1^{(1)*} = A_2^{(1)}A_2^{(1)*} = A_1^{(2)}A_1^{(2)*} = A_2^{(2)}A_2^{(2)*} = \tilde{I}$ ) were taken and the total small signal absorption at the frequency $\omega_1$ was set equal to $\sigma_1 N_0 \ell$ = 10 ( here $\ell$ is the medium length and $N_0$ is the total molecular density per unit volume). The main feature of the results obtained is that the coefficients $g_{2,1}^{(1)}$ and $g_{2,2}^{(2)}$ that are directly responsible for the interference interaction of counter-propagating waves have the sign opposite to the one of constant components $g_{0,1}^{(1)}$ and $g_{0,2}^{(2)}$ . Just therefore, in the general case the interference of counter-propagating waves causes either the reduction of radiation absorption (in absorbing media) or the decrease of gain (in active laser media). In particular, it follows from Fig.2 that these effects must be especially noticeable at the intensities close to saturation intensities of the corresponding transitions.

Figure 3 represents the comparison of the results of numerical integration of Eqs.(1.49), (1.50) and of the equations describing

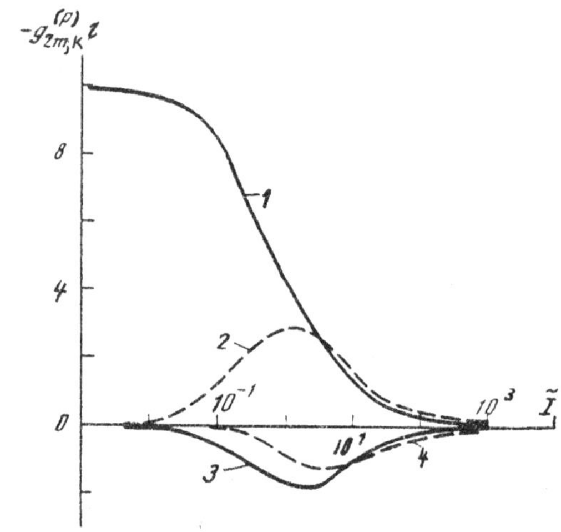


Fig. 2. Dependence of the two-wave interaction coefficients $g^{(p)}_{2m,k}l$ on the intensity $\tilde{I}$ in the three-level absorbing medium (the case of two-frequency radiation with widely separated frequencies $\omega_1$ and $\omega_2$ ) 1 - $g^{(1)}_{0,1}l$ ; 2 - $g^{(2)}_{0,2}l$ ; 3 - $g^{(1)}_{2,1}l$ ; 4 - $g^{(2)}_{2,2}l$ .

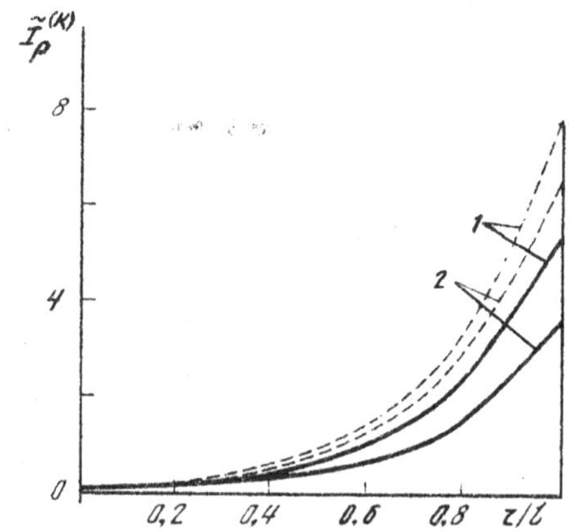


Fig. 3. Intensity distributions of forward waves $\tilde{I}^{(1)}_{1}$ (1) and $\tilde{I}^{(2)}_{1}$ (2) at the frequencies $\omega_1$ and $\omega_2$ calculated for the three-level system with inclusion (hevy lines) and without inclusion (dashed lines) of interference of counter-propagating waves.

amplification of counter-propagating waves without regard for interference interaction. The amplifying three-level system with $\tau_* = 10$, $\beta_{j1} = \beta_{j2} = 0.3$, $\alpha_1 = 0$, $\alpha_2 \doteq 1$, and $x_1 \sigma_1 N_0 \ell = 10$ (see (1.34), (1.35)) was considered. The input intensities of all the waves were set equal to $\tilde{I}_p^{(k)} = A_p^{(k)} A_p^{(k)*} = 0.1$.

It follows from Fig.3 that the wave intensities at the medium output decrease by a factor of 1.5-2 in this case if the interference effects are taken into account. The decrease in gain due to interference becomes more pronounced when the gain of the active medium is increased (with the growth of $\tau_*$ ).

### 1.3.2. Features of interference interaction of three-frequency radiation in multilevel resonant media

It follows from the previous sections that as the equations of interference interaction of counter-propagating waves are derived, the calculations become too cumbersome with increasing $N$ , the number of levels of the resonant medium. Therefore let us consider the case of degenerate two-wave interaction of three-frequency radiation in an amplifying multilevel medium as one more illustration.

Let us set $N = 3$ in Eqs.(1.15). As before, consider the excitation of molecules of the active medium in an electric discharge ( $W(t) = n_0/T_*$ ). Then the probabilities of excitation of separate levels are expressed by $\alpha_0 = -1$ and $\sum_{k=1}^{3} \alpha_k = 1$.

Introducing the designations

$$s_0 = 1 + \beta_{j_1} + \beta_{j_1}\beta_{j_2} + \beta_{j_1}\beta_{j_2}\beta_{j_3}, \quad \tau_* = T_1/T_*,$$
$$\tau_2 = T_1/T_2, \quad \tau_3 = T_1/T_3, \quad \sigma_2' = x_2\sigma_2/x_1\sigma_1, \quad \sigma_3' = x_3\sigma_3/x_1\sigma_1,$$

we obtain the following expressions for coefficients $f_p^{(k)}$ ( $k, p$ =

1, 2, 3) of set of equations (1.16)

$$
\begin{aligned}
f_1^{(1)} &= [\tau_*(\beta_{j_1}+\alpha_1)(1+\beta_{j_2}+\beta_{j_2}\beta_{j_3})-\tau_2\beta_{j_2}+(1+\beta_{j_1})]/s_0,\\
f_2^{(1)} &= [\tau_*(\beta_{j_1}+\alpha_1)(1+\beta_{j_3})-\tau_2(1+\beta_{j_1})-\beta_{j_1}(1+\beta_{j_1})(1+\beta_{j_3})]/(\sigma_2' s_0),\\
f_3^{(1)} &= [\tau_*(\beta_{j_1}+\alpha_1)+\tau_2\beta_{j_1}\beta_{j_2}-\beta_{j_1}(1+\beta_{j_1})]/(\sigma_3' s_0),\\
f_1^{(2)} &= (\sigma_2'/s_0)[\tau_*(\alpha_2-\beta_{j_2}\alpha_1)(1+\beta_{j_2}+\beta_{j_2}\beta_{j_3})+\\
&\quad+\tau_2\beta_{j_2}(1+\beta_{j_2})-\tau_3\beta_{j_2}\beta_{j_3}-\beta_{j_2}],\\
f_2^{(2)} &= [\tau_*(\alpha_2-\beta_{j_2}\alpha_1)+\tau_2(1+\beta_{j_1})(1+\beta_{j_2})-\tau_3\beta_{j_3}(1+\beta_{j_1})+\\
&\quad+\beta_{j_1}\beta_{j_2}(1+\beta_{j_3})]/s_0,\\
f_3^{(2)} &= (\sigma_2'/\sigma_3' s_0)[\tau_*(\alpha_2-\beta_{j_2}\alpha_1)-\tau_2\beta_{j_1}\beta_{j_2}(1+\beta_{j_2})-\\
&\quad-\tau_3(1+\beta_{j_1}+\beta_{j_1}\beta_{j_2})+\beta_{j_1}\beta_{j_2}],\\
f_1^{(3)} &= (\sigma_3'/s_0)[\tau_*(\alpha_3-\beta_{j_3}\alpha_2)(1+\beta_{j_2}+\beta_{j_2}\beta_{j_3})-\\
&\quad-\tau_2\beta_{j_2}\beta_{j_3}+\tau_3\beta_{j_2}\beta_{j_3}(1+\beta_{j_3})],\\
f_2^{(3)} &= (\sigma_3'/\sigma_2' s_0)[\tau_*(\alpha_3-\beta_{j_3}\alpha_2)(1+\beta_{j_3})-\tau_2\beta_{j_3}(1+\beta_{j_1})+\\
&\quad+\tau_3\beta_{j_3}(1+\beta_{j_1})(1+\beta_{j_3})],\\
f_3^{(3)} &= [\tau_*(\alpha_3-\beta_{j_3}\alpha_2)+\tau_2\beta_{j_1}\beta_{j_2}\beta_{j_3}+\tau_3(1+\beta_{j_3})(1+\beta_{j_1}+\beta_{j_1}\beta_{j_2})]/s_0.
\end{aligned}
$$

Then the coefficients $Q_p^{(k)}$ have the form

$$
\begin{aligned}
&Q_1^{(1)}=1,\ Q_2^{(1)}=-\tau_2/\sigma_2'(1+\beta_{j_2}),\ Q_3^{(1)}=0,\\
&Q_1^{(2)}=-\beta_{j_2}\sigma_2'/(1+\beta_{j_1}),\ Q_2^{(2)}=\tau_2,\ Q_3^{(2)}=-\sigma_2'\tau_3/\sigma_3'(1+\beta_{j_3}),\\
&Q_1^{(3)}=0,\ Q_2^{(3)}=-\tau_2\sigma_3'\beta_{j_3}/\sigma_2'(1+\beta_{j_2}),\quad Q_3^{(3)}=\tau_3,
\end{aligned}
$$

and the right sides ( $R_k$ ) are equal to

$$
\begin{aligned}
R_1 &= x_1\sigma_1 N_0[\tau_*(\beta_{j_1}+\alpha_1)+\tau_2\beta_{j_1}\beta_{i_2}-\beta_{j_1}(1+\beta_{j_1})]/s_0,\\
R_2 &= x_2\sigma_2 N_0[\tau_*(\alpha_2-\beta_{j_2}\alpha_1)-\tau_2\beta_{j_1}\beta_{j_2}(1+\beta_{j_2})+\\
&\quad+\tau_3\beta_{j_1}\beta_{j_2}\beta_{j_3}+\beta_{j_1}\beta_{j_2}]/s_0,\\
R_3 &= x_3\sigma_3 N_0[\tau_*(\alpha_3-\beta_{j_3}\alpha_2)+\tau_2\beta_{j_1}\beta_{j_2}\beta_{i_3}-\\
&\quad-\tau_3\beta_{j_1}\beta_{i_2}\beta_{j_3}(1+\beta_{j_3})]/s_0.
\end{aligned}
$$

Supposing that the lower molecular levels are satisfactorily described by the harmonic oscillator model, we shall also accept the relations $T_1/T_2 = \sigma_2/\sigma_1 = 2$ and $T_1/T_3 = \sigma_3/\sigma_1 = 3$.

In the steady-state mode, neglecting the diffusive spreading of spatial gratings formed by interference interaction ( i.e., for $D \to 0$) and using (1.23), we can write the equations for slowly varying wave amplitudes in the form

$$\frac{dA_1^{(p)}}{dz} = \frac{1}{2}\left(g_{0,\,p}^{(p)} A_1^{(p)} + g_{2,\,p}^{(p)} A_2^{(p)}\right),$$
$$-\frac{dA_2^{(p)}}{dz} = \frac{1}{2}\left(g_{2,\,p}^{(p)*} A_1^{(p)} + g_{0,\,p}^{(p)} A_2^{(p)}\right), \quad p = 1,\ 2,\ 3. \tag{1.51}$$

Determining the coefficients of two-wave interaction by the method used in section 1.2, we have

$$g_{2m,\,p}^{(p)} = S_p \lambda_p^m, \qquad p = 1,\ 2,\ 3, \qquad m = 0,\ 1,\ \ldots, \tag{1.52}$$

where

$$\lambda_p = -\frac{\xi_p}{2\eta_p C_{2,\,0}^{(p)*}}\left[1 - \left(1 - \frac{4\eta_p^2 C_{2,\,0}^{(p)} C_{2,\,0}^{(p)*}}{\xi_p^2}\right)^{1/2}\right].$$

Let us introduce the designations

$$\gamma_i^{(k)} = f_i^{(k)} + Q_i^{(k)} C_{0,\,0}^{(i)}, \qquad i,\ k = 1,\ 2,\ 3.$$

Then we can show that the coefficients $\xi_p$ in (1.52) are given by the expressions

$$\xi_1 = f_1^{(3)} + \frac{1}{d_1}\left[\gamma_2^{(3)}\left(f_3^{(1)}\gamma_1^{(2)} - \gamma_3^{(2)}\gamma_1^{(1)}\right) + \gamma_3^{(3)}\left(\gamma_2^{(2)}\gamma_1^{(1)} - \gamma_2^{(1)}\gamma_1^{(2)}\right)\right],$$
$$\xi_2 = \gamma_2^{(2)} + \frac{1}{d_2}\left[\gamma_1^{(2)}\left(f_3^{(1)}\gamma_2^{(3)} - \gamma_3^{(3)}\gamma_2^{(1)}\right) + \gamma_3^{(2)}\left(f_1^{(3)}\gamma_2^{(1)} - \gamma_1^{(1)}\gamma_2^{(3)}\right)\right],$$
$$\xi_3 = f_3^{(1)} + \frac{1}{d_3}\left[\gamma_1^{(1)}\left(\gamma_2^{(2)}\gamma_3^{(3)} - \gamma_2^{(3)}\gamma_3^{(2)}\right) + \gamma_2^{(1)}\left(f_1^{(3)}\gamma_3^{(2)} - \gamma_1^{(2)}\gamma_3^{(3)}\right)\right],$$

where

$$d_1 = \gamma_2^{(1)}\gamma_3^{(2)} - f_3^{(1)}\gamma_2^{(2)}, \qquad d_2 = \gamma_1^{(1)}\gamma_3^{(3)} - f_1^{(3)}f_3^{(1)},$$
$$d_3 = \gamma_1^{(2)}\gamma_2^{(3)} - f_1^{(3)}\gamma_2^{(2)}.$$

As regards the coefficients $\eta_p$ in (1.52), they have the form

$$\eta_1 = \frac{1}{d_1}\left\{\gamma_2^{(3)}\left(f_3^{(1)}Q_1^{(2)} - f_3^{(2)}Q_1^{(1)} - Q_3^{(2)}Q_1^{(1)}C_{0,\,0}^{(3)}\right) + \right.$$
$$\left. + \gamma_3^{(3)}\left[f_2^{(2)}Q_1^{(1)} - f_2^{(1)}Q_1^{(2)} + C_{0,\,0}^{(2)}\left(Q_2^{(2)}Q_1^{(1)} - Q_1^{(2)}Q_2^{(1)}\right)\right]\right\},$$

$$\eta_2 = Q_2^{(2)} + \frac{1}{d_2}[\gamma_1^{(1)}(f_3^{(1)}Q_2^{(3)} - \gamma_3^{(3)}Q_2^{(1)}) + \gamma_3^{(2)}(f_1^{(3)}Q_2^{(1)} - \gamma_1^{(1)}Q_2^{(3)})],$$

$$\eta_3 = \frac{1}{d_3}[\gamma_1^{(1)}(\gamma_2^{(2)}Q_3^{(3)} - \gamma_2^{(3)}Q_3^{(2)}) + \gamma_2^{(1)}(f_1^{(3)}Q_3^{(2)} - \gamma_1^{(2)}Q_3^{(3)})].$$

Finally, the arbitrary constants $S_p$, that are found from the boundary conditions, are the solutions of the set of three algebraic equations

$$\begin{aligned}
&(\gamma_1^{(1)} + Q_1^{(1)}Z_1)\, S_1 + (\gamma_2^{(1)} + Q_2^{(1)}Z_2)\, S_2 + f_3^{(1)}S_3 = R_1,\\
&(\gamma_1^{(2)} + Q_1^{(2)}Z_1)\, S_1 + (\gamma_2^{(2)} + Q_2^{(2)}Z_2)\, S_2 + (\gamma_3^{(2)} + Q_3^{(2)}Z_3)\, S_3 = R_2,\\
&f_1^{(3)}S_1 + (\gamma_2^{(3)} + Q_2^{(3)}Z_2)\, S_2 + (\gamma_3^{(3)} + Q_3^{(3)}Z_3)\, S_3 = R_3.
\end{aligned}$$

Here
$$Z_p = C_{2,0}^{(p)*}\lambda_p + C_{2,0}^{(p)}\lambda_p^*, \quad p = 1, 2, 3.$$

So, set of equations (1.51) with regard for (1.52) and (1.29) describes the stationary interference interaction of three-frequency radiation in the case of cascade transitions in four-level molecular media.

As far as analytical investigation of such equations seems to be impossible, we have used numerical methods for revealing the role of interference effects in the amplification of counter-propagating waves.

The numerical solution of the corresponding boundary-value problem was carried out by an iteration method using iterative integration of equations for forward and backward waves. Besides (1.51), we also integrated the equations

$$\frac{d\tilde{I}_1^{(p)}}{dz} = g_p\tilde{I}_1^{(p)}, \qquad -\frac{d\tilde{I}_2^{(p)}}{dz} = g_p\tilde{I}_2^{(p)}, \qquad p = 1, 2, 3,$$

that describe the amplification of counter-propagating waves in the absence of interference interaction (in this case the gain coeeficients $g_p$ are described by a set of algebraic equations following from (1.16) at $\partial/\partial t = 0$).

Some characteristic calculation results (for $\tau_*$ = 20, $\beta_{j1}$ =

$\beta_{j2} = \beta_{j3} = 0.3$, $\alpha_1 = \alpha_2 = 0$, $\alpha_3 = 1$, and $\tilde{I}_1^{(p)}(0) = \tilde{I}_2^{(p)}(\ell) = 0.1$) are illustrated in Figs.4 and 5.

It is interesting to note that the inclusion of interference effects in the media with a relatively low single-pass gain causes (as in the two-frequency case) only a reduction of the amplification level (see Fig.4) whereas in the media with a higher total gain a principally new phenomenon appears, an assymetry in the amplification of counter-propagating waves. In the case of symmetric boundary conditions it causes the appearance of multistable solutions of the problem under investigation.

Simulation of high-gain media show that the symmetric state characterizing, in particular, the distribution of intensities of bacward waves in the absence of interference (in Fig.6, the corresponding curves are depicted by dashed lines) appears to be unstable under the conditions of degenerate two-wave mixing and splits into two stable asymmetric states (heavy and dashed lines in Fig.5) to which the numerical iterative integration of set of equations (1.51) approaches.

It is clear that even small fluctuations of the active medium parameters (or, for example, temporal dephasing of counter-propagating waves) will inevitably cause the appearance of amplification assymmetry In particular, in the process of numerical integration of the equations of two-wave interaction, the disturbance of symmetry of the medium parameters was caused by iterative integration of the equations for wave amplitudes. Beginning the iterative process with integrating the equations for forward or backward waves, one can easily obtain both assymetric field distributions.

It is interesting to note that under certain conditions such an asymmetry arizes even in the case of a single-frequency two-wave interaction. So, the existence of asymmetric solutions (in the case of

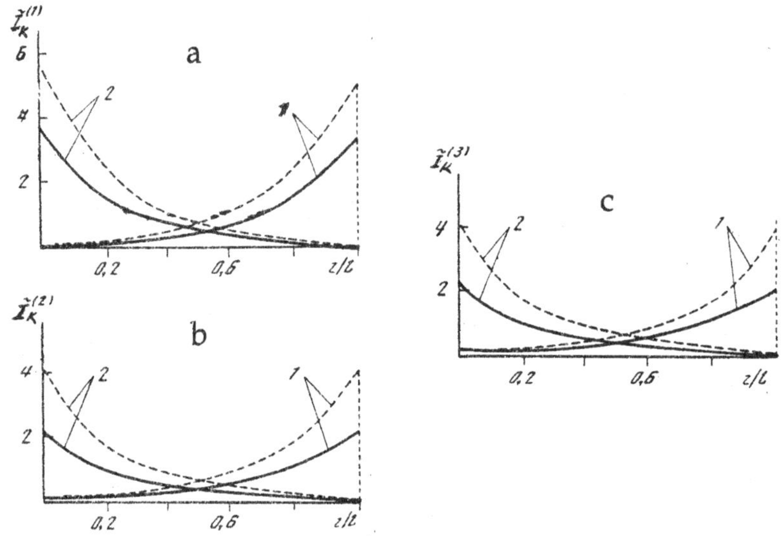


Fig. 4. Intensity distributions of forward (1) and backward (2) waves at the frequencies of transitions between the bands 1 ⟶ 0 (a), 2 ⟶ 1 (b), and 3 ⟶ 2 (c) in the four-level amplifying medium with inclusion (heavy lines) and without inclusion (dashed lines) of interference of counter-propagating waves ( $\varkappa_1 \sigma_1 N_0 l$ = 10).

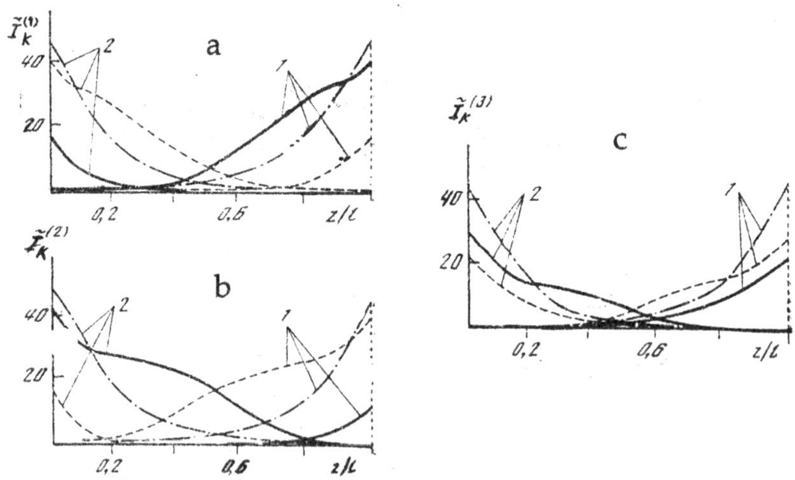


Fig. 5. Two possible asymmetric intensity distributions (heavy and dashed lines) of forward (1) and backward (2) waves at the frequencies of transitions between the bands 1 ⟶ 0 (a), 2 ⟶ 1 (b), and 3 ⟶ 2 (c) in the four-level high-gain ( $\varkappa_1 \sigma_1 N_0 l = 20$) amplifying medium with inclusion of interference effects. Dot-and-dash lines show intensity distributions without inclusion of interference effects.

symmetric boundary conditions) is shown in [3] for the equations of interference interaction in the presence of linear damping in the amplifying medium.

## 1.4 On degenerate two-wave interaction of multifrequency radiation in the case of close frequencies

The previous analysis concerned the case when interference of counter-propagating waves with different frequencies could be neglected. As it was already noted in section 1.2, such a situation takes place if the frequency detuning $\Delta_{k,p}$ between the frequencies $\omega_k$, $\omega_p$ is large and the interference effects in the interaction of waves with different frequencies are of little importance. But if the conditions for this approximation (for more details see section 1.2) are not fulfilled, then the general picture of interference interaction becomes more complicated and requires a more detailed analysis [37].

To illustrate the problems that arise in this case, let us dwell on a relatively simple example of the two-frequency two-level interaction considered earlier in section 1.3.1.

Setting $N = 2$, let us represent the gain coefficients $g_k$ on the transition between the $1 \longrightarrow 0$ and $2 \longrightarrow 1$ bands in the form of expansions (1.22). Then

$$g_k(z, t) = g_{0,k}^{(k)} + \sum_{m=1}^{\infty} [(g_{2m,k}^{(1)} e^{-i2mq^{(1)}z} + \text{c. c.}) + (g_{2m,k}^{(2)} e^{-i2mq^{(2)}z} + \text{c. c.})], \quad k = 1, 2. \tag{1.53}$$

Let us introduce the frequency detuning $\Delta = q^{(1)} - q^{(2)} > 0$ (we suppose that $\omega_1 > \omega_2$) and then address to Eqs. (1.20) in order to obtain the equations of two-wave interaction for our case.

Taking into account that $q^{(2)} = q^{(1)} - \Delta$, we substitute (1.53) into (1.20). Then, collecting the terms with identical factors of the form $\exp(\pm i q^{(1)} r)$ in the left and right sides of this equation, we obtain the equations for slowly varying field amplitudes $A_p^{(1)}$ at the frequency $\omega_1$:

$$\frac{\partial A_1^{(1)}}{\partial z} + \frac{1}{c}\frac{\partial A_1^{(1)}}{\partial t} = \frac{1}{2}\left[g_{0,1}^{(1)} A_1^{(1)} + \left(g_{2,1}^{(1)} + g_{2,1}^{(2)} e^{i2\Delta z}\right) A_2^{(1)}\right],$$
$$-\frac{\partial A_2^{(1)}}{\partial z} + \frac{1}{c}\frac{\partial A_2^{(1)}}{\partial t} = \frac{1}{2}\left[\left(g_{2,1}^{(1)*} + g_{2,1}^{(2)*} e^{-i2\Delta z}\right) A_1^{(1)} + g_{0,1}^{(1)} A_2^{(1)}\right]. \tag{1.54}$$

In the same way we obtain the equations for the field at the frequency $\omega_2$:

$$\frac{\partial A_1^{(2)}}{\partial z} + \frac{1}{c}\frac{\partial A_1^{(2)}}{\partial t} = \frac{1}{2}\left[g_{0,2}^{(2)} A_1^{(2)} + \left(g_{2,2}^{(2)} + g_{2,2}^{(1)} e^{-i2\Delta z}\right) A_2^{(2)}\right],$$
$$-\frac{\partial A_2^{(2)}}{\partial z} + \frac{1}{c}\frac{\partial A_2^{(2)}}{\partial t} = \frac{1}{2}\left[\left(g_{2,2}^{(2)*} + g_{2,2}^{(1)*} e^{i2\Delta z}\right) A_1^{(2)} + g_{0,2}^{(2)} A_2^{(2)}\right]. \tag{1.55}$$

Deriving the equations for coefficients $g_{2m,k}^{(p)}$ on the basis of (1.16), (1.21), and (1.53), we take account of the relation between $q^{(1)}$, $q^{(2)}$, and $\Delta$. Introducing the designations

$$Y_{0,k} = g_{0,k}^{(k)}, \qquad Y_{m,k} = g_{2m,k}^{(1)} + g_{2m,k}^{(2)} e^{i2m\Delta z}, \tag{1.56}$$

we can represent the sought equations in the form

$$\frac{\partial Y_{0,k}}{\partial t} + \sum_{p=1}^{2}\left\{f_p^{(k)} Y_{0,p} + Q_p^{(k)}\left[C_{0,0}^{(p)} Y_{0,p} + C_{2,0}^{(p)} Y_{1,p}^{*} e^{i(p-1)2\Delta z} + C_{2,0}^{(p)*} Y_{1,p} e^{-i(p-1)2\Delta z}\right]\right\} = R_k,$$
$$\frac{\partial Y_{m,k}}{\partial t} + \frac{1}{\tau_{2m,D}^{(1)}} g_{2m,k}^{(1)} + \frac{1}{\tau_{2m,D}^{(2)}} g_{2m,k}^{(2)} e^{i2m\Delta z} + \sum_{p=1}^{2}\left\{f_p^{(k)} Y_{m,p} + Q_p^{(k)}\left[C_{0,0}^{(p)} Y_{m,p} + C_{2,0}^{(p)} Y_{m-1,p} e^{i(p-1)2\Delta z} + C_{2,0}^{(p)*} Y_{m+1,p} e^{-i(p-1)2\Delta z}\right]\right\} = 0, \quad k = 1, 2; \quad m = 1, 2, 3, \ldots \tag{1.57}$$

The quantities $1/\tau_{2m,D}^{(1)} = 4(q^{(1)})^2 Dm^2$ and $1/\tau_{2m,D}^{(2)} = 4(q^{(2)})^2 Dm^2$ in Eqs. (1.57) are the characteristic times of diffusive spreading of the $m$ th-order interference gratins at the frequencies $\omega_1$ and $\omega_2$, respectively.

The second equation in (1.57) shows that the presence of diffusive terms in Eqs. (1.16) that account for diffusive motion of molecules ( $D \neq 0$), causes in the end not only diffusive spreading but also mixing up of the spatial gratings formed by the fields with different frequencies (because $\tau_{k,D}^{(1)} \neq \tau_{k,D}^{(2)}$ in the general case). The latter means that the use of expansion $g_k(z,t)$ in the form (1.53) in the case under consideration is, generally speaking, incorrect because set of equations (1.57) is not closed with respect to the variables $Y_{m,k}$.

It is obvious that in the case of widely separated frequencies, those of the terms in (1.57) that contain rapidly oscillating factors of the form $\exp(\pm i2m\Delta z)$ may be omitted. Then a closed set of equations with respect to coefficients $g_{2m,k}^{(p)}$ can be obtained; such equations were examined in the previous section.

In another extreme case of very close frequencies, one can suppose that $\tau_{k,D}^{(1)} \simeq \tau_{k,D}^{(2)} \simeq \tau_{k,D}$. Therefore set of equations (1.57) may be considered closed again because

$$\frac{1}{\tau_{2m,D}^{(1)}} g_{2m,k}^{(1)} + \frac{1}{\tau_{2m,D}^{(2)}} g_{2m,k}^{(2)} e^{i2m\Delta z} \simeq \frac{1}{\tau_{2m,D}} Y_{m,k}.$$

Further, if the vibrational relaxation time of molecules is shorter than the time of diffusive spreading of the gratings, then the diffusive term in (1.57) may be omitted. In this latter case, set of equations (1.57) also appears to be closed. In particular, the characteristic values of diffusive coefficients for gases at atmospheric pressure are $D \gtrsim 0.1$ cm$^2$/s and for the wavelength $\lambda \sim 3$–$10$ μm we have $1/q^2D \sim 10^{-6} - 10^{-7}$ s. Therefore, if the relaxation times satisfy the relation $T_k \lesssim 10^{-6} - 10^{-7}$ s, then the role of diffusive spreading of the gratings is unimportant.

Neglecting the diffusive spreading of the gratings (i.e., supposing formally that $D \to 0$), let us go on to the examination of a steady-state mode of two-frequency interference interaction in the case of close frequencies $\omega_1$ and $\omega_2$. In the steady-state case, the following chain of closed equations for $Y_{m,k}$ is obtained from (1.57)

$$\begin{aligned}&(f_1^{(1)}+Q_1^{(1)}C_{0,0}^{(1)})\,Y_{m,1}+Q_1^{(1)}C_{2,0}^{(1)}Y_{m-1,1}+Q_1^{(1)}C_{2,0}^{(1)*}Y_{m+1,1}=\\&=-\{(f_2^{(1)}+Q_2^{(1)}C_{0,0}^{(2)})\,Y_{m,2}+Q_2^{(1)}C_{2,0}^{(2)}Y_{m-1,2}e^{i2\Delta z}+\\&+Q_2^{(1)}C_{2,0}^{(2)*}Y_{m+1,2}e^{-i2\Delta z}\},\\&(f_1^{(2)}+Q_1^{(2)}C_{0,0}^{(1)})\,Y_{m,1}+Q_1^{(2)}C_{2,0}^{(1)}Y_{m-1,1}+Q_1^{(2)}C_{2,0}^{(1)*}Y_{m+1,1}=\\&=-\{(f_2^{(2)}+Q_2^{(2)}C_{0,0}^{(2)})\,Y_{m,2}+Q_2^{(2)}C_{2,0}^{(2)}Y_{m-1,2}e^{i2\Delta z}+\\&+Q_2^{(2)}C_{2,0}^{(2)*}Y_{m+1,2}e^{-i2\Delta z}\},\qquad m=1,\ 2,\ 3,\ \dots\end{aligned} \tag{1.58}$$

The boundary conditions for set (1.58) are found from the first equation of (1.57):

$$\begin{aligned}&(f_1^{(1)}+Q_1^{(1)}C_{0,0}^{(1)})\,Y_{0,1}+Q_1^{(1)}\,(C_{2,0}^{(1)}Y_{1,1}^{*}+C_{2,0}^{(1)*}Y_{1,1})+\\&+(f_2^{(1)}+Q_2^{(1)}C_{0,0}^{(2)})\,Y_{0,2}+Q_2^{(1)}\,(C_{2,0}^{(2)}Y_{1,2}^{*}e^{i2\Delta z}+C_{2,0}^{(2)*}Y_{1,2}e^{-i2\Delta z})=R_1,\\&(f_1^{(2)}+Q_1^{(2)}C_{0,0}^{(1)})\,Y_{0,1}+Q_1^{(2)}\,(C_{2,0}^{(1)}Y_{1,1}^{*}+C_{2,0}^{(1)*}Y_{1,1})+\\&+(f_2^{(2)}+Q_2^{(2)}C_{0,0}^{(2)})\,Y_{0,2}+Q_2^{(2)}\,(C_{2,0}^{(2)}Y_{1,2}^{*}e^{i2\Delta z}+C_{2,0}^{(2)*}Y_{1,2}e^{-i2\Delta z})=R_2.\end{aligned} \tag{1.59}$$

The solution of Eqs. (1.58) is sought in the form

$$Y_{m,1}=S\lambda^m,\qquad Y_{m,2}=\varkappa Y_{m,1},\qquad m=0,\ 1,\ \dots, \tag{1.60}$$

where $S$ and $\varkappa$ are some constants.

Substituting (1.60) into (1.58), we obtain two following expressions for $\varkappa$:

$$\begin{aligned}\varkappa&=-\frac{Q_1^{(1)}C_{2,0}^{(1)}+(f_1^{(1)}+Q_1^{(1)}C_{0,0}^{(1)})\,\lambda+Q_1^{(1)}C_{2,0}^{(1)*}\lambda^2}{Q_2^{(1)}C_{2,0}^{(2)}e^{i2\Delta z}+(f_2^{(1)}+Q_2^{(1)}C_{0,0}^{(2)})\,\lambda+Q_2^{(1)}C_{2,0}^{(2)*}e^{-i2\Delta z}\lambda^2},\\ \varkappa&=-\frac{Q_1^{(2)}C_{2,0}^{(1)}+(f_1^{(2)}+Q_1^{(2)}C_{0,0}^{(1)})\,\lambda+Q_1^{(2)}C_{2,0}^{(1)*}\lambda^2}{Q_2^{(2)}C_{2,0}^{(2)}e^{i2\Delta z}+(f_2^{(2)}+Q_2^{(2)}C_{0,0}^{(2)})\,\lambda+Q_2^{(2)}C_{2,0}^{(2)*}e^{-i2\Delta z}\lambda^2}.\end{aligned} \tag{1.61}$$

Since $\varkappa$ is assumed to be a constant, we equate these two expressions. This gives us the equation of the fourth degree (with complex coefficients) for $\lambda$:

$$b_0\lambda^4+b_1\lambda^3+b_2\lambda^2+b_3\lambda+b_4=0, \tag{1.62}$$

where

$$b_0 = \beta C_{2,0}^{(1)*} C_{2,0}^{(2)*} e^{-i2\Delta z},$$
$$b_1 = (p + \beta C_{0,0}^{(1)}) C_{2,0}^{(2)*} e^{-i2\Delta z} + (q + \beta C_{0,0}^{(2)}) C_{2,0}^{(1)*},$$
$$b_2 = \beta (C_{2,0}^{(1)} C_{2,0}^{(2)*} e^{-i2\Delta z} + C_{2,0}^{(1)*} C_{2,0}^{(2)} e^{i2\Delta z}) + + \beta C_{0,0}^{(1)} C_{0,0}^{(2)} + q C_{0,0}^{(1)} + p C_{0,0}^{(2)} + \gamma,$$

$b_3 = b_1^*$, $b_4 = b_0^*$, where

$$\beta = Q_1^{(1)} Q_2^{(2)} - Q_1^{(2)} Q_2^{(1)}, \quad \gamma = f_1^{(1)} f_2^{(2)} - f_1^{(2)} f_2^{(1)},$$
$$p = Q_2^{(2)} f_1^{(1)} - Q_2^{(1)} f_1^{(2)}, \quad q = Q_1^{(1)} f_2^{(2)} - Q_1^{(2)} f_2^{(1)}.$$

The complex roots of the fourth order equation can be found by standard methods.

By the peculiar features of coefficients it turns out that two roots ( $\lambda_1$, $\lambda_2$ ) of the four are inside the circle $|\lambda| \leq 1$, while the other two ( $\lambda_3$, $\lambda_4$ ) are outside this region. Being interested in the limited solution of infinite set (1.58), we neglect the growing solutions with $|\lambda_3| > 1$ and $|\lambda_4| > 1$. As a result, the solution of set (1.58) can be represented in the form

$$Y_{m,1} = S_1 \lambda_1^m + S_2 \lambda_2^m, \quad Y_{m,2} = S_1 \varkappa_1 \lambda_1^m + S_2 \varkappa_2 \lambda_2^m, \tag{1.63}$$

Here $m$ = 0, 1, 2, ..., and $\varkappa_j = \varkappa(\lambda_j)$, where $\varkappa(\lambda_j)$ is determined by any of the formulas (1.61).

The arbitrary constants in (1.63) are determined from boundary conditions (1.59). One can easily note that in the present case $S_1$

and $\delta_2$ appear to be complex, therefore actually the two equations in (1.59) are reduced to a system of four linear algebraic equations for the real and imaginary parts of $\delta_1$ and $\delta_2$.

As an example let us consider the passive absorbing medium that was already examined before. Its gain coefficients $g_1$ and $g_2$ are determined by relations (1.37). Let the intensity $\tilde{I}_k^{(p)}$ ( $k, p$ = 1, 2) of all the four waves incident on the medium be equal to the saturation intensities of corresponding transitions ( $\tilde{I}_k^{(p)}$ = 1). Then, suppose that the small signal absorption of the medium at the frequency $\omega_1$ ( 0 → 1 transition) is $\sigma_1 N_0 \ell$ = 10. If the interference of counter-propagating waves is neglected, then in these conditions we obtain from (1.37) the steady-state values of absorption coefficients at 0 → 1 and 1 → 2 transitions $(-g_1\ell)$ = $(-g_2\ell)$ = $2/7\,\sigma_1 N_0 \ell \simeq 2.86$.

Now let us account for the interference interaction of counter--propagating waves at close frequencies $\omega_1$ and $\omega_2$. the roots of Eq.(1.62) were determined by the Descartes-Euler method (see [38]) and their dependence on the phase shift $\varphi = 2\Delta Z$ caused by wave detuning is depicted in Fig.6 (it shows the roots lying inside the circle $|\lambda| \leq 1$).

The analogous dependences for coefficients of two-wave interaction $g_{0,1}^{(1)}\ell$, $g_{0,2}^{(2)}\ell$, $(g_{2,1}^{(1)}\ell + g_{2,1}^{(2)}e^{i2\Delta Z})\ell$, and $(g_{2,2}^{(2)} + g_{2,2}^{(1)}e^{-i2\Delta Z})\ell$ (that enter into equations for wave amplitudes (1.54), (1.55)) calculated from the values of these roots (using formulas (1.63)) are depicted in Fig.7. One can see that the constant components $g_{0,1}^{(1)}$ and $g_{0,2}^{(2)}$ are in fact (as expected) real quantities (arg ( $g_{0,k}^{(k)}$ ) = 0), but they appreciably differ from both the stationary values $g_1$ and $g_2$ in the absence of interference effects and the values $g_{0,1}^{(1)}$ and $g_{0,2}^{(2)}$ obtained earlier neglecting the interference interaction of counter-propagating waves with different frequencies (cf. with Fig.2). We note that within the wave detuning period (0, $2\pi$), the pulsations of all the coefficients of two-wave interaction are observed. In principle, the latter can

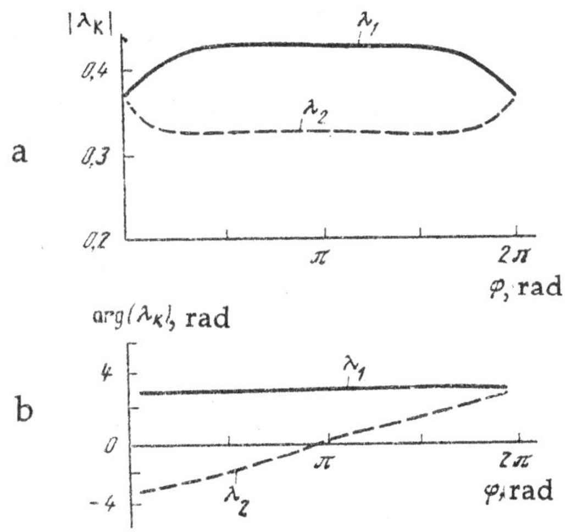


Fig. 6. Dependences of the modulus (a) and phase (b) of complex roots $\lambda_1$ and $\lambda_2$ on the wave detuning $\varphi = 2\Delta z$ in the absorbing resonant medium at $\tilde{\sigma}_1 N_0 \ell = 10$:

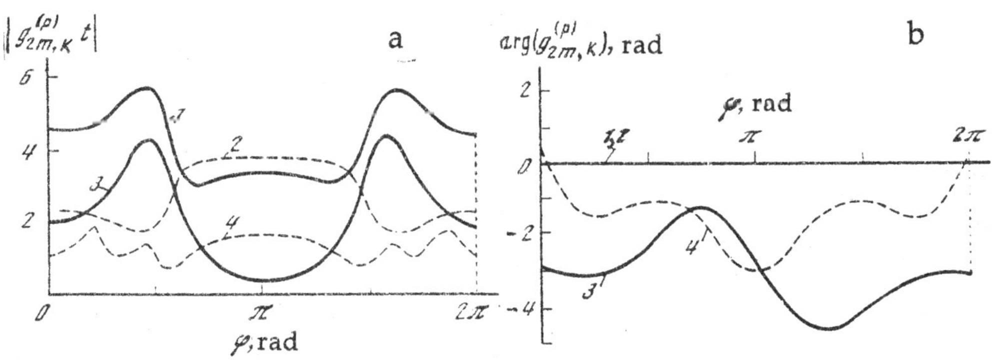


Fig. 7. Dependences of the modulus (a) and phase (b) of complex coefficients $g^{(p)}_{2m,k}\ell$ of two-wave interaction on the wave detuning at $\tilde{\sigma}_1 N_0 \ell = 10$

$1 - g^{(1)}_{0,1}\ell$; $2 - g^{(2)}_{0,2}\ell$; $3 - (g^{(1)}_{2,1} + g^{(2)}_{2,1}e^{i2\Delta z})\ell$; $4 - (g^{(2)}_{2,2} + g^{(1)}_{2,2}e^{-i2\Delta}$

give rise to oscillations of the interacting wave intensities.

So, the analysis performed shows that in the case of close frequencies $\omega_1$ and $\omega_2$, the exact algorithm for calculating the interference interaction of counter-propagating waves exists too. However, such calculations are essentially far from trivial because determination of two-wave interaction coefficients in every space point needs a rather cumbersome procedure. It is also clear that in the general case of multifrequency radiation the situation becomes much more complicated.

## 2. DEGENERATE FOUR-WAVE INTERACTION (DFI) OF IR RADIATION IN RESONANT MEDIA

An interest in degenerate four-wave interaction is caused, first of all, by the possibility of realizing PC of low-intensity optical signals (in contrast to SBS and SRS, for example) in a zero-threshold mode. It must be also noted that DFI is essentially the only effective method of PC in the middle- and far-IR ranges at present [7,25]. We remind that the DFI method is based on generation of a conjugated wave in the process of mixing three waves with equal frequencies in a nonlinear medium. Various nonlinear media may be used for this purpose, however at high-energy levels, low absorbing liquids and resonant gases are of the greatest interest, as a rule. The two main nonlinearity mechanisms that show themselves in the latter media are the resonant (saturable) absorption (amplification) of radiation [11] and also the thermal mechanism caused by light-induced heat deposition in absorbing and amplifying media (see, for example, [4-7,39]). A large number of publications is devoted to theoretical and experimental investigation of these DFI mechanisms. Besides the above-cited references, we note, in particular, [40-50] and also [51-52] (the latter two consider DFI in gaseous media with a chemical thermal nonlinearity).

In the following section, we shall consider a relatively simple case of a two-level absorbing (amplifying) medium with the resonant and the thermal nonlinearity mechanisms and obtain the equations describing nonstationary DFI of counter-propagating waves at arbitrary field intensities of the interacting waves. And further we shall investigate a more complicated case of multifrequency DFI for cascade transitions in multilevel molecular media.

## 2.1 Equations for nonstationary DFI of counter-propagating waves in the media with resonant and thermal nonlinearity at arbitrary field intensities

Consider DFI of counter-propagating waves in the medium containing a buffer gas (diluent) and a certain part of "two-level" particles (atoms or molecules) that resonantly interact with the radiation field at the frequency $\omega$ (Fig.8). We suppose that the content of diluent in the mixture is sufficiently large, so that radiative heating is not too high and therefore constant characteristic parameters of the gas medium may be taken.

Let us denote the coefficient of small signal absorption (for intensity) by resonant particles by $\alpha_0$ . We also suppose for generality that the medium may contain some nonresonant admixtures characterized by the linear absorption coefficient $\alpha_1$ .

Taking this into account, let us write the wave equation for the radiation field in the form

$$\nabla^2 E = \frac{\alpha_1}{c}\frac{\partial E}{\partial t} + \frac{1}{c^2}\frac{\partial^2 E}{\partial t^2} + \frac{4\pi}{c^2}\frac{\partial^2 P}{\partial t^2}, \qquad (2.1)$$

where the first term in the right part phenomenologically accounts for nonresonant losses in the medium; $P = P_0 + P_1$ is polarization consisting of nonresonant ( $P_0$ ) and resonant ( $P_1$ ) parts; $E = \sum_{j=1}^{4} E_j =$

$\sum_{j=1}^{4} ( \mathcal{E}_j e^{i\omega t} + c.c. )$ the total electromagnetic field of the four interacting waves.

Let us denote the electromagnetic field amplitudes of the reference waves (directed strictly in the opposite directions) by $E_1$ and $E_2$, and the amplitudes of signal and conjugated waves by $E_3$ and $E_4$, respectively. The angle between the directions of the waves $E_1$ and $E_3$ is denoted by $\theta$ (we suppose that $\theta < \pi$).

Using the nonresonant refractive index of the medium $n$, let us take account of the fact that the modulated component of refractive index (caused by the temperature and the density modulation) is relatively small, as a rule, so that $n = n_0 + \tilde{n}$, where $\tilde{n} \ll n_0$. Since $E + 4\pi P_0 = n^2 E$, we have $E + 4\pi P_0 \simeq ( n_0^2 + 2 n_0 \tilde{n} ) E$.

As for the resonant component of polarization $P_1 = \mathcal{P} e^{i\omega t} + c.c.$, then in a quasistatic (with respect to the field) approximation, one can use for it relation (1.4). As a result, representing the spatial dependences of the fields in the form $\mathcal{E}_j = A_j e^{-ik_j r}$, we reduce (by analogy with (1.5)) the wave equation (2.1) to the form

$$\sum_{j=1}^{4} \left( \mathbf{k}_j \frac{\partial A_j}{\partial \mathbf{r}} + \frac{k}{c} \frac{\partial A_j}{\partial t} \right) e^{-i\mathbf{k}_j \mathbf{r}} = -ik^2 \left\{ \frac{\tilde{n}}{n_0} - \frac{1}{2k} [i\alpha_1 + (i + \delta) g] \right\} \sum_{j=1}^{4} A_j e^{-i\mathbf{k}_j \mathbf{r}}, \tag{2.2}$$

where $g$ is the frequency-dependent absorption coefficient of the two-level medium and $\delta$ is the frequency detuning.

Neglecting the rapid sound oscillations of density in the subsequent analysis, we suppose that the main variations of refractive index are caused by relatively slow temperature effects. Then

$$n\,(T) \simeq n\,(T_0) + (\partial n / \partial T)_p\,(T - T_0), \quad \tilde{n} \simeq n(T) - n(T_0), \tag{2.3}$$

where $T_0$ is the initial temperature and the derivative $\partial n / \partial T$ is taken at constant pressure because of the thermal process inertia.

Since $(\partial n/\partial T)_p \simeq -(n_0 - 1)/T$ for the gases, then taking account of (2.3) we can rewrite Eq.(2.2) in the form (here and further the wave nonstationarity will be neglected for simplicity)

$$\sum_{j=1}^{4} \frac{k_j}{k}\frac{\partial A_j}{\partial r} e^{-ik_j r} = -\left[\frac{1}{2}(\alpha_1 + \gamma_0 g) + i\gamma_1(T - T_0)\right]\sum_{j=1}^{4} A_j e^{-ik_j r}, \qquad (2.4)$$

where $\gamma_0 = 1 - i\delta$ and $\gamma_1 = k(n_0 - 1)/T_0$. We also took into account that $n_0 \simeq 1$ in the gases.

For the self-consistent description of DFI, Eq.(2.4) must be supplemented with the equations describing variation of $g$ and $T$ in the medium under the action of radiation field.

The equations for absorption coefficient of the two-level medium follow directly from (1.2), where just the spatial diffusion of excited particles should be taken into account. Then

$$\frac{\partial g}{\partial t} + \frac{1}{\tau_1}(g - \alpha_0) = -2\sigma g I + D\nabla^2 g. \qquad (2.5)$$

Here $\alpha_0 = \sigma N$ is the coefficient of small signal absorption of the two-level medium (we suppose that the initial poulation of the upper level is practically equal to zero); $N$ the volume concentration of resonant particles; $\sigma$ the frequency-dependent cross section of the transition considered; $D$ the diffusion coefficient, and $\tau_1$ the relaxation time of the upper level.

To describe the temperature effects, let us turn to the equation of conservation of energy in this system. The equation of conservation of energy for the motionless medium with the inclusion of thermal conduction, linear light absorption by the admixtures (thermalization process is supposed to be practically instantaneous), and resonant interaction of the radiation field with the medium can be taken in the form

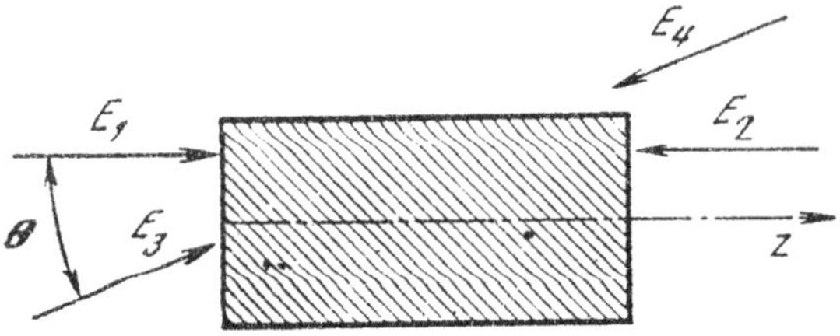


Fig. 8. Simplified configuration of four-wave interaction in a resonant medium.

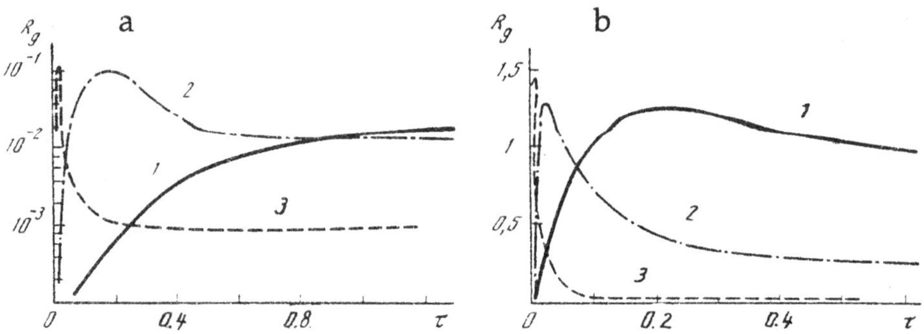


Fig. 9. Time dependences of the dynamical reflection coefficient $R_g$ in the absorbing (a), $\alpha_0 l$ = -4 and amplifying (b), $\alpha_0 l$ = 4 media at $\tilde{I}_3^0 / \tilde{I}_1^0 = 10^{-3}$:

a - $\tilde{I}_1^0 / \tilde{I}_s = \tilde{I}_2^l / \tilde{I}_s$ = 1 (curve 1), 10 (curve 2), and $10^2$ (curve 3);

b - $\tilde{I}_1^0 / \tilde{I}_s = \tilde{I}_2^l / \tilde{I}_s$ = 0.1 (curve 1), 1 (curve 2), and 10 (curve 3).

$$\rho \frac{\partial h}{\partial t} - \frac{\partial p}{\partial t} = (\alpha_1 + g) I + \lambda \nabla^2 T. \tag{2.6}$$

Here $\rho$ is the mixture density; $p$ the pressure; $\lambda$ the thermal conductivity coefficient; $h = \sum_k \rho_k h_k/\rho$ the specific enthalpy of the mixture and $h_k = H_k/W_k$ , where $H_k = H_k^\circ + C_{p_k} T$ is the molar enthalpy of the $k$ th component, the molecular weight ( $H_k^\circ$ the combination heat, $C_{p_k}$ the molar heat capacity of the $k$ th-component at constant pressure). We note that the radiation intensity in (2.6) is expressed in units of energy.

The chosen form of the equation of conservation of energy is convenient, first of all, in the situations requiring due regard for various kinetic and chemical transformations, in particular, in the media with a thermal chemical nonlinearity.

In the present case, chemical transformations are absent in the medium, the molecular weight does not vary and the two-level medium is considered. Therefore, introducing the molar particle concentration for the corresponding levels $n_k = N_k/N_A$ ( $N_A$ is the Avogadro number) and performing simple transformations, we obtain from (2.6)

$$\frac{1}{W} \rho c_V \frac{\partial T}{\partial t} - \lambda \nabla^2 T = - \sum_{k=1}^{2} H_k^0 \frac{\partial n_k}{\partial t} + (\alpha_1 + g) I, \tag{2.7}$$

where $C_V$ is the molar heat capacity of the mixture at constant volume.

$H_2^\circ - H_1^\circ = N_A h\nu$ ( $h\nu$ is the radiation quantum energy).

Therefore, taking into consideration the equations for $N_k$ :

$$\partial N_1/\partial t = N_2/\tau_1 - gI + D\,\nabla^2 N_1,$$
$$\partial N_2/\partial t = -N_2/\tau_1 + gI + D\,\nabla^2 N_2$$

(the fist level is supposed to be the ground one), we reduce (2.7) to the form

$$\frac{1}{W}\rho c_V \frac{\partial T}{\partial t} = \alpha_1 I - \frac{h\nu}{2\sigma\tau_1}(g-\alpha_0) + \lambda\nabla^2 T - \frac{h\nu}{2\sigma} D\nabla^2 g. \tag{2.8}$$

Taking into account that the saturation intensity of the two-level system (in quanta) is $I_s = 1/2\,\sigma\tau_1$, we normalize the wave fields setting $e_j = \mathcal{E}_j\sqrt{I_s}$, $a_j = A_j/\sqrt{I_s}$, and $\tilde{I} = I/I_s$. As a result, Eqs. (2.4), (2.5), and (2.8) can be written in the form

$$\begin{aligned} &\sum_{j=1}^{4}\frac{1}{k}\mathbf{k}_j\frac{\partial a_j}{\partial \mathbf{r}}e^{-i\mathbf{k}_j\mathbf{r}} = -\left[\frac{1}{2}(\alpha_1+\gamma_0 g) - i\gamma_1(T-T_0)\right]\sum_{j=1}^{4}a_j e^{-i\mathbf{k}_j\mathbf{r}},\\ &\frac{\partial g}{\partial t} + \frac{1}{\tau_1}(g-\alpha_0) = -\frac{1}{\tau_1}g\tilde{I} + D\nabla^2 g,\\ &\frac{\partial T}{\partial t} = \frac{\beta}{\tau_1}(\alpha_1\tilde{I} - g + \alpha_0) + \varkappa\nabla^2 T - \beta D\nabla^2 g, \end{aligned} \tag{2.9}$$

where $\beta = Wh\nu/2\rho c_V\sigma$ and $\varkappa = W\lambda/\rho c_V$ is the temperature conductivity coefficient of the medium.

Using relations (2.9), one can get the set of equations describing nonstationary DFI of counter-propagating waves.

Since the phase matching condition for DFI of counter-propagating waves has the form $\mathbf{k}_1 + \mathbf{k}_2 = \mathbf{k}_3 = \mathbf{k}_4 = 0$ and the total radiation intensity (with account of interference effects) is $\tilde{I} = ee^*$, where $e = \sum_{j=1}^{4} a_j e^{-i\mathbf{k}_j\mathbf{r}}$, then the expression for $\tilde{I}$ can be represented in the form

$$\begin{aligned} \tilde{I} = C_{0,0} + (C_{2,0}e^{-i2\mathbf{k}_1\mathbf{r}} + C_{0,2}e^{-i2\mathbf{k}_3\mathbf{r}} +\\ + C_{1,1}e^{-i\mathbf{k}_1\mathbf{r}-i\mathbf{k}_3\mathbf{r}} + \tilde{C}_{1,1}e^{-i\mathbf{k}_1\mathbf{r}+i\mathbf{k}_3\mathbf{r}} + \text{c. c.}. \end{aligned} \tag{2.10}$$

Here we used the notations

$$C_{0,0}=\sum_{j=1}^{4}a_j a_j^*,\quad C_{2,0}=a_1a_2^*,\quad C_{0,2}=a_3a_4^*,$$
$$C_{1,1}=a_1a_4^*+a_2^*a_3,\quad \tilde{C}_{1,1}=a_1a_3^*+a_2^*a_4 \tag{2.11}$$

On account of (2.9)-(2.11), the quantities $g(\mathbf{r},t)$ and $T(\mathbf{r},t)$ should be sought in the form of the following expansions in the series in spatial harmonics of the interacting waves ( $x = g$ or $T$ ) [53-55]:

$$x(r,t)=x_{0,0}+\sum_{m=1}^{\infty}\sum_{l=0}^{2m}(x_{2m-l,l}\exp\{-i(2m-l)\mathbf{k}_1\mathbf{r}-il\mathbf{k}_3\mathbf{r}\}+$$
$$+\tilde{x}_{2m-l,l}\exp\{-i(2m-l)\mathbf{k}_1\mathbf{r}-il\mathbf{k}_3\mathbf{r}\}+\text{c. c.} \tag{2.12}$$

We note that the quantities $C_{p,l}$ and $x_{p,l}$ in (2.10)-(2.12) are the symmetrized, while $\tilde{C}_{p,l}$ and $\tilde{x}_{p,l}$ are the antisymmetrized coefficients of the expansions (with respect to the signs of the terms $p\mathbf{k}_1\mathbf{r}$ and $l\mathbf{k}_3\mathbf{r}$ in an exponent). As for the quantities $C_{2,0}$, $x_{2,0}$ and $C_{0,2}$, $x_{0,2}$, they are obviously responsible for the two-wave interaction of counter-propagating waves ( $E_1$, $E_2$ ) and ( $E_3$, $E_4$ ), respectively.

Directing the $z$ axis along the diagonal between the directions of propagation of the waves $E_1$ and $E_3$, let us set $\zeta = z/\cos(\theta/2)$. Then, using (2.12), we substitute the expansions of $g(\mathbf{r},t)$ and $T(\mathbf{r},t)$ into the first equation of (2.9) and collect the terms with similar rapidly varying factors in the left and right sides of it. As a result, we find the set of bound equations for slowly varying complex wave amplitudes:

$$\partial a_1/\partial\zeta = -{}^1\!/_2\gamma_0(g_{0,0}a_1+g_{2,0}a_2+\tilde{g}_{1,1}a_3+\tilde{g}_{1,1}a_4)+$$
$$+i\gamma_1[(T_{0,0}-T_0)a_1+T_{2,0}a_2+\tilde{T}_{1,1}a_3+T_{1,1}a_4]-{}^1\!/_2\alpha_1a_1,$$
$$-\partial a_2/\partial\zeta = -{}^1\!/_2\gamma_0(g_{2,0}^*a_1+g_{0,0}a_2+g_{1,1}^*a_3+\tilde{g}_{1,1}^*a_4)+$$
$$+i\gamma_1[T_{2,0}^*a_1+(T_{0,0}-T_0)a_2+T_{1,1}^*a_3+\tilde{T}_{1,1}^*a_4]-{}^1\!/_2\alpha_1a_2, \tag{2.13}$$
$$\partial a_3/\partial\zeta = -{}^1\!/_2\gamma_0(\tilde{g}_{1,1}^*a_1+g_{1,1}a_2+g_{0,0}a_3+g_{0,2}a_4)+$$
$$+i\gamma_1[\tilde{T}_{1,1}^*a_1+T_{1,1}a_2+(T_{0,0}-T_0)a_3+T_{0,2}a_4-{}^1\!/_2\alpha_1a_3,$$

$$-\partial a_4/\partial\zeta = -{}^1/_2\gamma_0(g^*_{1,1}a_1 + \tilde{g}_{1,1}a_2 + g^*_{0,2}a_3 + g_{0,0}a_4) +$$
$$+ i\gamma_1[T^*_{1,1}a_1 + \tilde{T}_{1,1}a_2 + T^*_{0,2}a_3(T_{0,0} - T_0)a_4] - {}^1/_2\alpha_1 a_4.$$

In the same way we obtain from the second equation of (2.9) an infinite chain of bound equations for coefficients $g_{p,l}$ ($p$, $l$ = 0, 1, 2, ...). In particular, the equations for the first five coefficients $g_{p,l}$ that directly enter into the right sides of set (2.13), have the form

$$\frac{\partial g_{0,0}}{\partial t} + \frac{1}{\tau_1}(g_{0,0} - \alpha_0) = -\frac{1}{\tau_1}[C_{0,0}g_{0,0} + (C_{2,0}g^*_{2,0} +$$
$$+ C_{0,2}g^*_{0,2} + C_{1,1}g^*_{1,1} + \tilde{C}_{1,1}\tilde{g}^*_{1,1} + \text{c.c.})],$$
$$\frac{\partial g_{2,0}}{\partial t} + \left(\frac{1}{\tau_1} + f_{2,0}\right)g_{2,0} = -\frac{1}{\tau_1}(C_{0,0}g_{2,0} + C_{2,0}g_{0,0} + C^*_{2,0}g_{4,0} +$$
$$+ C_{0,2}\tilde{g}_{2,2} + C^*_{0,2}g_{2,2} + C_{1,1}\tilde{g}_{1,} + C^*_{1,1}g_{3,1} + \tilde{C}_{1,1}g_{1,1} + \tilde{C}^*_{1,1}\tilde{g}_{3,1}),$$
$$\frac{\partial g_{0,2}}{\partial t} + \left(\frac{1}{\tau_1} + f_{0,2}\right)g_{0,2} = -\frac{1}{\tau_1}(C_{0,0}g_{0,2} + C_{2,0}\tilde{g}^*_{2,2} + C^*_{2,0}g_{2,2} +$$
$$+ C_{0,2}g_{0,0} + C^*_{0,2}g_{0,4} + C^*_{1,1}\tilde{g}^*_{1,1} + C^*_{1,1}g_{1,3} + \tilde{C}_{1,1}\tilde{g}^*_{1,3} + \tilde{C}^*_{1,1}g_{1,1}), \tag{2.14}$$
$$\frac{\partial g_{1,1}}{\partial t} + \left(\frac{1}{\tau_1} + f_{1,1}\right)g_{1,1} = -\frac{1}{\tau_1}(C_{0,0}g_{1,1} + C_{2,0}\tilde{g}^*_{1,1} + C^*_{2,0}g_{3,1} +$$
$$+ C_{0,2}\tilde{g}_{1,1} + C^*_{0,2}g_{1,3} + C_{1,1}g_{0,0} + C^*_{1,1}g_{2,2} + \tilde{C}_{1,1}g_{0,2} + \tilde{C}^*_{1,1}g_{2,0}),$$
$$\frac{\partial \tilde{g}_{1,1}}{\partial t} + \left(\frac{1}{\tau_1} + \tilde{f}_{1,1}\right)\tilde{g}_{1,1} = -\frac{1}{\tau_1}(C_{0,0}\tilde{g}_{1,1} + C_{2,0}g^*_{1,1} + C^*_{2,0}\tilde{g}_{3,1} +$$
$$+ C_{0,2}\tilde{g}_{1,3} + C^*_{0,2}g_{1,1} + C_{1,1}g^*_{0,2} + C^*_{1,1}g_{2,0} + \tilde{C}_{1,1}g_{0,0} + \tilde{C}^*_{1,1}\tilde{g}_{2,2}].$$

The following designations are used in (2.14):

$$f_{p,l} = k^2D\left[(p+l)^2\cos^2\frac{\tilde{\theta}}{2} + (p-l)^2\sin^2\frac{\tilde{\theta}}{2}\right],$$
$$\tilde{f}_{p,l} = k^2D\left[(p-l)^2\cos^2\frac{\tilde{\theta}}{2} + (p+l)^2\sin^2\frac{\tilde{\theta}}{2}\right], \tag{2.15}$$

where $\tilde{\theta} = \theta$ for $p$, $l \geq 1$ and $\tilde{\theta} = 0$ if $p = 0$ or $l = 0$. It is obvious that the quantity $\tau_D = 1/k^2D$ is the characteristic time of diffusive spreading of the amplitude gratings.

We note that the equations for higher order coefficients $g_{p,l}$ and $\tilde{g}_{p,l}$ have the form analogous to (2.14) and their right sides always contain nine terms with nine different coefficients $C_{0,0}$, $C_{2,0}$, $C^*_{2,0}$, ..., $\tilde{C}^*_{1,1}$ as one of the two factors. And the second factor is given by various coefficients $g_{p,l}$ and $\tilde{g}_{p,l}$ or their complex conjugates. The concrete form of the latter factors is determined by the requirement that the index ( $p$ or $l$ ) of the coefficient $g_{p,l}$ standing in the left side should match the algebraic sum of the same type indices of the corresponding factors (with due account of symmetrization and the sign of complex conjugation). As a result, the terms with identical exponents are separated out in the second equation of (2.9).

The equations for coefficients $T_{p,l}$ and $\tilde{T}_{p,l}$ entering into the right sides of set (2.13) are obtained from the third equation of (2.9):

$$\begin{aligned}
&\frac{\partial T_{0,0}}{\partial t} = \frac{\beta\alpha_1}{\tau_1} C_{0,0} - \frac{\beta}{\tau_1}(g_{0,0} - \alpha_0),\\
&\frac{\partial T_{2,0}}{\partial t} + d_{2,0} T_{2,0} = -\beta f_{2,0} g_{2,0} + \frac{\beta}{\tau_1}\alpha_1 C_{2,0} - g_{2,0}),\\
&\frac{\partial T_{0,2}}{\partial t} + d_{0,2} T_{0,2} = -\beta f_{0,2} g_{0,2} + \frac{\beta}{\tau_1}(\alpha_1 C_{0,2} - g_{0,2}),\\
&\frac{\partial T_{1,1}}{\partial t} + d_{1,1} T_{1,1} = -\beta f_{1,1} g_{1,1} + \frac{\beta}{\tau_1}(\alpha_1 C_{1,1} - g_{1,1}),\\
&\frac{\partial \tilde{T}_{1,1}}{\partial t} + \tilde{d}_{1,1} \tilde{T}_{1,1} = -\beta \tilde{f}_{1,1} \tilde{g}_{1,1} + \frac{\beta}{\tau_1}(\alpha_1 \tilde{C}_{1,1} - \tilde{g}_{1,1}),
\end{aligned} \tag{2.16}$$

where

$$\begin{aligned}
&d_{p,l} = k^2\varkappa\left[(p+l)^2\cos^2\frac{\tilde{\theta}}{2} + (p-l)^2\sin^2\frac{\tilde{\theta}}{2}\right],\\
&\tilde{d}_{p,l} = k^2\varkappa\left[(p-l)^2\cos^2\frac{\tilde{\theta}}{2} + (p+l)^2\sin^2\frac{\tilde{\theta}}{2}\right],
\end{aligned} \tag{2.17}$$

and $\tau_\varkappa = 1/k^2\varkappa$ is, obviously, the characteristic time of thermal spreading of phase gratings.

So, in the general case set of equations (2.13), (2.14), and (2.16) describing nonstationary DFI of counter-propagating waves in the media with resonant and thermal nonlinearities appears to be rather cumbersome. However, the characteristic times of diffusive and thermal spreading of many amplitude and phase gratings are extremely short and one can take advantage of this property. Indeed, most of the gases have $D \sim \varkappa \geq 0.1\ cm^2/s$ at room temperature and atmospheric pressure. Consequently, we have $\tau_D \sim \tau_\varkappa \lesssim 10^{-9}-10^{-7}$ s for optical waves in the range 1–10 µm. The times $\tau_D$ and $\tau_\varkappa$ shorten with decreasing the gas pressure.

If the angle between the waves $E_1$ and $E_3$ is small ($\theta \ll 1$), then according to (2.15) and (2.17) the long-lived gratings correspond to the coefficients $\tilde{g}_{m,m}$ and $\tilde{T}_{m,m}$ and the characteristic times of diffusive and thermal spreading are of the order of $1/k^2m^2D\theta^2$ and $1/k^2m^2\varkappa\theta^2$, respectively. If $\theta$ is small and $m$ is not very large, these times may far exceed the population relaxation time $\tau_1$. So, it is evident from (2.14) and (2.16) that for the durations of radiation pulses $t_p \gg \tau_D, \tau_\varkappa$ one can neglect the long-lived amplitude and phase gratings in the actual conditions because the corresponding coefficients $g_{p,l}$ and $T_{p,l}$ are very small. In this case the set of DFI equations is essentially simplified.

In particular, the equations for wave amplitudes give

$$
\begin{aligned}
\partial a_1/\partial\zeta &= -{}^1\!/_2\gamma_0(g_{0,0}a_1 + \tilde{g}_{1,1}a_3) + i\gamma_1[(T_{0,0} - T_0)a_1 + \tilde{T}_{1,1}a_3] - {}^1\!/_2\alpha_1 a_1,\\
-\partial a_2/\partial\zeta &= -{}^1\!/_2\gamma_0(g_{0,0}a_2 + \tilde{g}^*_{1,1}a_4) + i\gamma_1[(T_{0,0} - T_0)a_2 + \tilde{T}^*_{1,1}a_4] - {}^1\!/_2\alpha_1 a_2,\\
\partial a_3/\partial\zeta &= -{}^1\!/_2\gamma_0(g_{0,0}a_3 + \tilde{g}^*_{1,1}a_1) + i\gamma_1[(T_{0,0} - T_0)a_3 + \tilde{T}^*_{1,1}a_1] - {}^1\!/_2\alpha_1 a_3,\\
-\partial a_4/\partial\zeta &= -{}^1\!/_2\gamma_0(g_{0,0}a_4 + \tilde{g}_{1,1}a_2) + i\gamma_1[(T_{0,0} - T_0)a_4 + \tilde{T}_{1,1}a_2] - {}^1\!/_2\alpha_1 a_4,
\end{aligned}
\tag{2.18}
$$

The infinite chain of equations for coefficients $\tilde{g}_{m,m}$ obtains the form

$$
\begin{aligned}
&\frac{\partial g_{0,0}}{\partial t} + \frac{1}{\tau_1}(g_{0,0} - \alpha_0) = -\frac{1}{\tau_1}(C_{0,0}g_{0,0} + \tilde{C}_{1,1}\tilde{g}^*_{1,1} + \tilde{C}^*_{1,1}\tilde{g}_{1,1}),\\
&\frac{\partial \tilde{g}_{m,m}}{\partial t} + \left(\frac{1}{\tau_1} + \tilde{f}_{m,m}\right)\tilde{g}_{m,m} =\\
&= -\frac{1}{\tau_1}(C_{0,0}\tilde{g}_{m,m} + \tilde{C}_{1,1}\tilde{g}_{m-1,m-1} + \tilde{C}^*_{1,1}\tilde{g}_{m+1,m+1}), \qquad m = 1, 2, 3, \ldots
\end{aligned}
\tag{2.19}
$$

Finally, the set of equations for $T_{p,l}$ gives

$$\frac{\partial T_{0,0}}{\partial t} = \frac{\beta}{\tau_1}\left[\alpha_1 C_{0,0} - (g_{0,0} - \alpha_0)\right], \tag{2.20}$$

$$\frac{\partial \tilde{T}_{1,1}}{\partial t} + \tilde{d}_{1,1}\tilde{T}_{1,1} = \beta\left[-\left(\frac{1}{\tau_1} + \tilde{f}_{1,1}\right)\tilde{g}_{1,1} + \frac{\alpha_1}{\tau_1}\tilde{C}_{1,1}\right].$$

Set of equations (2.18)-(2.20) obtained describes nonstationary DFI of counter-propagating waves in the media with resonant and thermal nonlinearity at arbitrary field intensities but sufficiently small interaction angles $\theta \ll 1$.

## 2.2. The features of nonstationary DFI in the media with resonant nonlinearity

As it was first noted in [56,57], the four-wave interaction of short optical pulses in resonant media essentially differs from DFI under steady-state conditions because of the nonstationarity of the processes. In particular, it was experimentally shown that if the radiation pulses are shorter than the characteristic time of population relaxation of a resonant medium, then using DFI it is possible to obtain reflection coefficients of the signal wave that considerably exceed the values obtained under steady-state conditions. Some features of DFI of short optical pulses were examined in [16,17,53,54].

Let us examine such features of nonstationary DFI in resonant media using the equations derived in the previous section. As a rule, the contribution of inertial thermal nonlinearity to DFI for the case of short optical pulses (in comparison with the population relaxation time of the two-level system $\tau_1$) is small. Therefore we set $\gamma_1 = 0$ in Eqs. (2.18). Besides, we suppose that nonresonant absorption is also absent in the medium ( $\alpha_1 = 0$). Then, restricting ourselves for simplicity to the small signal approximation ( $|a_3|, |a_4| \ll |a_1|, |a_2|$ ) and neglecting the effects associated with the finite time of wave propagation, we obtain from (2.18) the equations

$$\begin{aligned} \partial a_3/\partial\zeta &= -{}^1/_2\gamma_0\,(g_{0,0}\,a_3 + \tilde{g}^*_{1,1}a_1), \\ -\partial a_4/\partial\zeta &= -{}^1/_2\gamma_0\,(g_{0,0}a_4 + \tilde{g}_{1,1}a_2). \end{aligned} \tag{2.21}$$

In view of (2.19), the coefficients of four-wave interaction are given by the relations (we note that the wave amplitudes and their intensities in Eqs. (2.18)-(2.20) and below are normalized to the saturation intensity of the two-level medium)

$$\begin{aligned} \frac{\partial g_{0,0}}{\partial t} + \frac{1}{\tau_1}(1 + C_{0,0})\,g_{0,0} &= \frac{\alpha_0}{\tau_1}, \\ \frac{\partial \tilde{g}_{1,1}}{\partial t} + \left(\tilde{f}_{1,1} + \frac{1 + C_{0,0}}{\tau_1}\right)\tilde{g}_{1,1} &= -\frac{1}{\tau_1}\,g_{0,0}\tilde{C}_{1,1}. \end{aligned} \tag{2.22}$$

Since the temporal variation of strong reference waves $a_1$ and $a_2$ can be neglected, we find from (2.22) in the first approximation (for simplicity of calculations, $\tilde{f}_{1,1} \ll 1/\tau_1$ is taken)

$$\begin{aligned} g_{0,0} &= \frac{\alpha_0}{1 + C_{0,0}}(1 + C_{0,0}e^{-t/\tau}), \\ \tilde{g}_{1,1} &= -\frac{\alpha_0 C_{0,0}C_{1,1}}{1 + C_{0,0}}\,\frac{t}{\tau_1}\,e^{-t/\tau_1} - \frac{\alpha_0 C_{1,1}}{(1 + C_{0,0})^2}(1 - e^{-t/\tau}), \end{aligned} \tag{2.23}$$

where

$$\tau = \tau_1/(1 + C_{0,0}).$$

Considering $t \to \infty$ in (2.23), we obtain the steady-state value of coefficient of four-wave interaction

$$(\tilde{g}_{1,1})_{\text{CT}} = -\frac{\alpha_0 C_{1,1}}{(1 + C_{0,0})^2}. \tag{2.24}$$

On the other hand, determining the extremum of $\tilde{g}_{1,1}(t)$, we get

$$(\tilde{g}_{1,1})_{\max} = -\frac{\alpha_0 C_{1,1}}{1 + C_{0,0}}\,e^{-\frac{t_m}{\tau}} - \frac{\alpha_0 C_{1,1}}{(1 + C_{0,0})^2}\left(1 - e^{-\frac{t_m}{\tau}}\right), \tag{2.25}$$

where $t_m = \tau_1 / C_{0,0}$.

It appears from Eqs. (2.21) that the efficiency of nonstationary DFI in the strong fields is practically entirely determined by the temporal behavior of the coefficient $\tilde{g}_{1,1}$. Since we have $t_m/\tau \sim 1$ in the strong fields ( $C_{0,0} \gg 1$), it follows from the expression for $(\tilde{g}_{1,1})_{max}$ that the extreme values of $\tilde{g}_{1,1}$ may far exceed their steady-state values. This means that in principle considerably higher reflection coefficients of the signal wave may be obtained in the case of short optical pulses in comparison with the steady-state conditions. One can easily see that the pulse durations $t_p$ at which nonstationary DFI appears to be more effective than the nonstationary one, must not exceed the characteristic time $t_m$ in order of magnitude, where $t_m \simeq \tau_r = 1/\sigma I$ is the time of radiative saturation of the resonant medium.

It should be noted that in the weak fields ( $C_{0,0} \ll 1$) we always have $t_m/\tau = (1 + C_{0,0})/C_{0,0} \gg 1$. Therefore in the weak fields, the efficiency of nonstationary DFI does not exceed the efficiency of stationary interaction because $\tilde{g}_{1,1}(t) \lesssim (\tilde{g}_{1,1})_{st}$ in this case.

Direct numerical integration of the equations of DFI in the media with resonant (saturable) nonlinearity clearly confirms the conclusions of the analysis performed. So, Figs. 9a, b illustrate the results of the calculation of dynamical reflection coefficient $R_g = I_4^0 / I_3^0$ (here and below $\tilde{I}_k^{(0)} = \tilde{I}_k(0)$ and $\tilde{I}_k^{\ell} = I_k(\ell)$) obtained by integrating complete set of equations (2.13) (for $\gamma_1 = \alpha_1 = 0$) and the chain of equations for coefficients $g_{p,\ell}$ with the inclusion of the expansion terms up to the third approximation ( $p + \ell \leq 6$) in (2.12). The exact resonance was considered ( $\delta = 0$, $\gamma_0 = 1$) and, besides, diffusive spreading of the amplitude gratings was neglected ( $f_{p,\ell} = \tilde{f}_{p,\ell} = 0$). It should be noted that the main result of the inclusion of diffusive motion of the particles is the reduction of the stationary values of $R$. Hence the nonstationary effects in DFI become more apparent [54].

In amplifying media (Fig. 9b), the intensities of interacting waves appreciably increase during their propagation. Therefore, effective nonstationary DFI is observed in this case even at moderate intensities of the reference waves at the input of the medium.

## 2.3. Equations of stationary DFI in the media with resonant and thermal nonlinearity

Returning to set of equations (2.18)–(2.20), let us suppose that

$$a_j = b_j \exp\left[\pm i\gamma_1 (T_{0,0} - T_0)\zeta\right], \tag{2.26}$$

The plus sign in (2.26) corresponds to $j$ = 1 and 3 and the minus sign corresponds to $j$ = 2 and 4, respectively.

Substituting (2.26) into (2.18), we eliminate in the latter set the terms determining the nonstationary phase shift caused by variation of the constant component of refractive index in time. As a result, taking into account that $C_{0,0} = \sum_{j=1}^{4} b_j b_j^*$ and $\tilde{C}_{1,1} = b_1 b_3^* + b_2^* b_4$, we obtain from (2.18)–(2.20) for $t \to \infty$ the equations for stationary DFI of counter-propagating waves in the media with resonant and thermal nonlinearities.

Eqs. (2.18) give

$$\begin{aligned}
db_1/d\zeta &= -{}^1\!/_2\gamma_0 (g_{0,0} b_1 + \tilde{g}_{1,1} b_3) + i\gamma_1 \tilde{T}_{1,1} b_3 - {}^1\!/_2\alpha_1 b_1, \\
-db_2/d\zeta &= -{}^1\!/_2\gamma_0 (g_{0,0} b_2 + \tilde{g}^*_{1,1} b_4) + i\gamma_1 \tilde{T}^*_{1,1} b_4 - {}^1\!/_2\alpha_1 b_2, \\
db_3/d\zeta &= -{}^1\!/_2\gamma_0 (g_{0,0} b_3 + \tilde{g}^*_{1,1} b_1) + i\gamma_1 \tilde{T}^*_{1,1} b_1 - {}^1\!/_2\alpha_1 b_3, \\
-db_4/d\zeta &= -{}^1\!/_2\gamma_0 (g_{0,0} b_4 + \tilde{g}_{1,1} b_2) + i\gamma_1 \tilde{T}_{1,1} b_2 - {}^1\!/_2\alpha_1 b_4,
\end{aligned} \tag{2.27}$$

where on account of (2.20)

$$\tilde{T}_{1,1} = \frac{\beta}{d_{1,1}}\left[\frac{\alpha_1}{\tau_1}\tilde{C}_{1,1} - \left(\frac{1}{\tau_1} + \tilde{f}_{1,1}\right)\tilde{g}_{1,1}\right]. \qquad (2.28)$$

In the case being considered, set of equations (2.19) for coefficients $\tilde{g}_{m,m}$ is reduced to the form

$$\begin{aligned}&(1 + C_{0,0})\, g_{0,0} + \tilde{C}_{1,1}\tilde{g}^*_{1,1} + \tilde{C}^*_{1,1}\tilde{g}_{1,1} = \alpha_0,\\ &(1 + C_{0,0})\, \tilde{g}_{m,m} + \tilde{C}_{1,1}\tilde{g}_{m-1,m-1} + \tilde{C}^*_{1,1}\tilde{g}_{m+1,m+1} = \tau_1\tilde{f}_{m,m}\tilde{g}_{m,m},\\ &m = 1, 2, 3, \ldots\end{aligned} \qquad (2.29)$$

In principle, it has an exact solution only when the quantities $\tilde{f}_{m,m}$ do not depend on $m$.

We note that at small interaction angles one can always fulfil the condition $\tau_1\tilde{f}_{m,m} \ll 1$ by choosing $\theta$ at least for small $m$ values. On the other hand, the coefficients of four-wave interaction rapidly decrease in modulus with increasing $m$. Therefore in the strong fields ( $C_{0,0} \gg 1$), the effect of the right hand side in the second equation of (2.29) on the character of solution may become essential only in the region of large $m$. Since the description of DFI needs nothing more than the knowledge of coefficients $g_{0,0}$ and $\tilde{g}_{1,1}$, one can use their values obtained from the exact solution of set (2.29) at $\tilde{f}_{m,m} = 0$ as a good approximation for this purpose:

$$g_{0,0} = \frac{\alpha_0}{\Delta}, \qquad \tilde{g}_{1,1} = -\frac{\alpha_0}{2\tilde{C}^*_{1,1}\Delta}(1 + C_{0,0} - \Delta), \qquad (2.30)$$

where $\Delta = [(1 + C_{0,0})^2 - 4\tilde{C}_{1,1}\tilde{C}^*_{1,1}]^{1/2}$.

In view of (2.28) and (2.29), we reduce set of equations (2.27) to the closed form

$$\begin{aligned}db_1/d\zeta &= -{}^1/_2\alpha_1 b_1 - \gamma_0 p\,(b_1 - \Psi b_3) + i\gamma_1 Q\,(\alpha_1\tilde{C}_{1,1} + p\Psi)\, b_3,\\ -db_2/d\zeta &= -{}^1/_2\alpha_1 b_2 - \gamma_0 p\,(b_2 - \Psi^* b_4) + i\gamma_1 Q\,(\alpha_1\tilde{C}^*_{1,1} + p\Psi^*)\, b_4,\end{aligned} \qquad (2.31)$$

$$db_3/d\zeta = -{}^1/_2\alpha_1 b_3 - \gamma_0 p\,(b_3 - \Psi^* b_1) + i\gamma_1 Q\,(\alpha_1 \tilde{C}^*_{1,1} + p\Psi^*)\, b_1,$$

$$-db_4/d\zeta = -{}^1/_2\alpha_1 b_4 - \gamma_0 p\,(b_4 - \Psi b_2) + i\gamma_1 Q\,(\alpha_1 \tilde{C}_{1,1} + p\Psi)\, b_2,$$

where $p = \alpha_0/2\Delta$, $Q = \beta/\tau_1 \tilde{d}_{1,1}$ and $\Psi = (1 + C_{0,0} - \Delta)/2\tilde{C}^*_{1,1}$.

Now let us consider in more detail some consequencies following from the equations derived.

### 2.3.1. Stationary DFI in the media with resonant nonlinearity

For simplicity, let us neglect in (2.31) the effects associated with the thermal nonlinearity. Setting $\alpha_1 = \gamma_1 = 0$, we obtain

$$\frac{db_1}{d\zeta} = -\frac{\gamma_0\alpha_0}{2\Delta}(b_1 - \Psi b_3), \qquad -\frac{db_2}{d\zeta} = -\frac{\gamma_0\alpha_0}{2\Delta}(b_2 - \Psi^* b_4),$$
$$\frac{db_3}{d\zeta} = -\frac{\gamma_0\alpha_0}{2\Delta}(b_3 - \Psi^* b_1), \qquad -\frac{db_4}{d\zeta} = -\frac{\gamma_0\alpha_0}{2\Delta}(b_4 - \Psi b_2). \tag{2.32}$$

Set of equations (2.32) describes stationary DFI in the counter--propagation configuration in the media with resonant nonlinearity with the inclusion of the diffusion of gas particles. It is clear from the derivation of these equations that they are valid only for sufficiently small angles of convergence $\theta$ (more exactly, the application domain of Eqs.(2.32) is bounded by the condition $\tau_1 \tilde{f}_{1,1} \ll 1$).

Figures 10 and 11 illustrate some results obtained by numerical integration of these equations for the case of absorbing medium at exact resonance ($\gamma_0 = 1$). So, Figs.10a, b show the dependences of the reflection coefficient of the signal wave $R = \tilde{I}_4(0)/\tilde{I}_3(0)$ on the intensity level of the reference wave at $\tilde{I}_2(\ell) = \tilde{I}_1(0)$ and $\tilde{I}_3(0)/\tilde{I}_1(0) = 10^{-3}$. Fig.10a illustrates the results of the standard

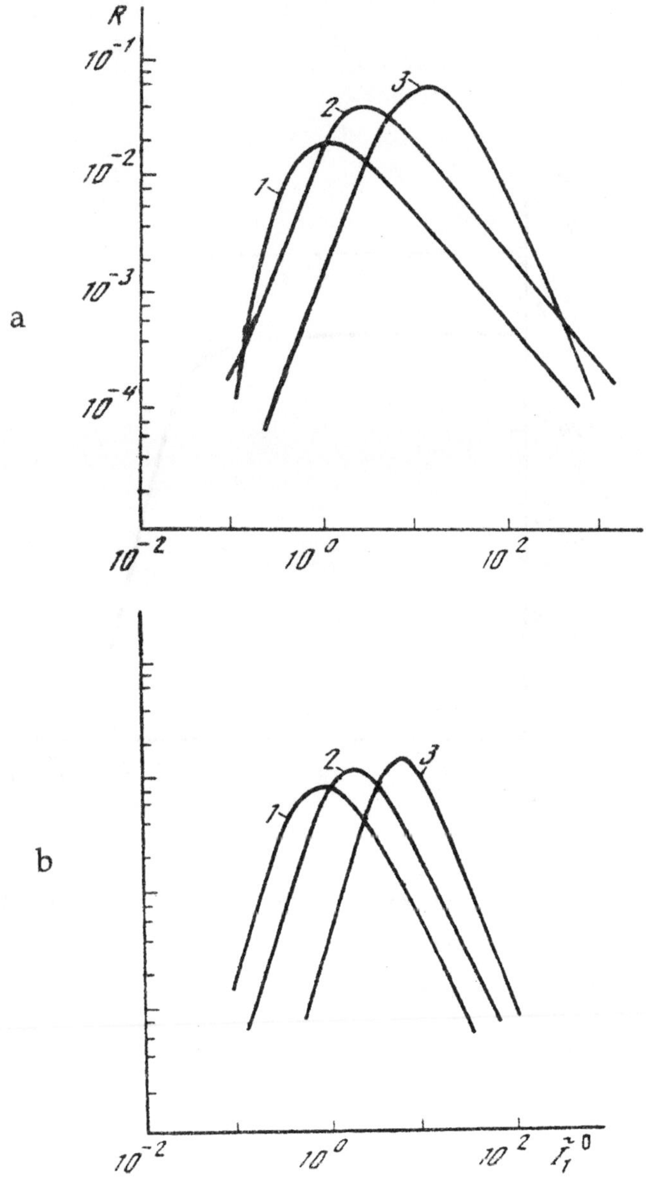


Fig. 10. Dependence of the signal wave reflection coefficient $R$ on the reference wave intensity $\tilde{I}_1^0$ in the absorbing resonant medium with inclusion (a) and without inclusion (b) of diffusive spreading of amplitude gratings ( $\alpha_0 \ell$ = 2 (1), 4 (2), and 8 (3)).

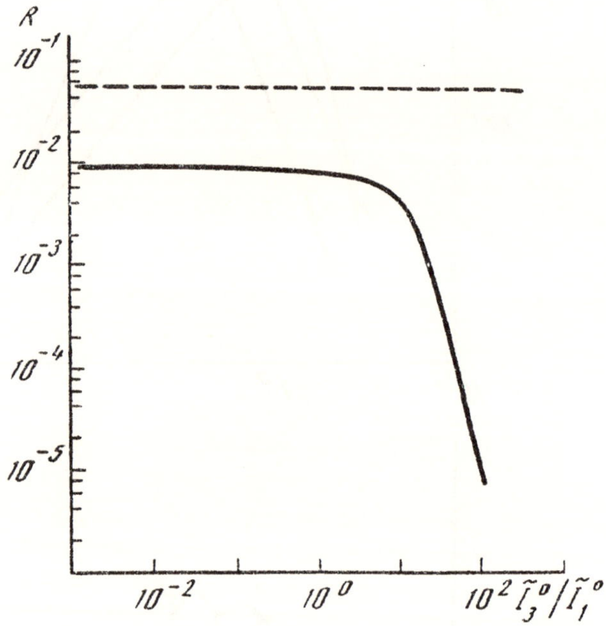


Fig. 11. Effect of the signal wave intensity on the reflection efficiency in the resonant absorbing medium ( $\tilde{I}_1^0$ = 10).
The dashed line illustrates the results of small-signal theory.

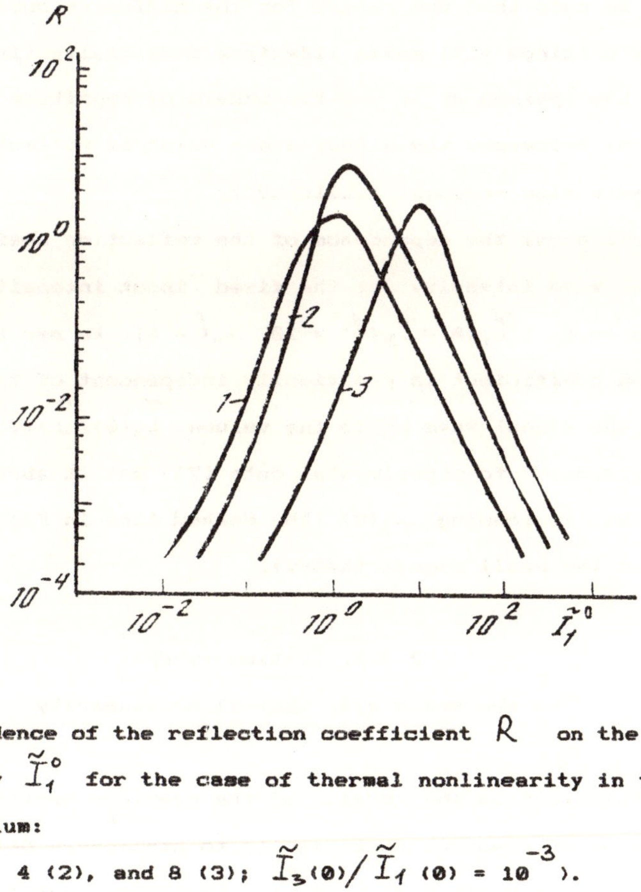


Fig. 12. Dependence of the reflection coefficient $R$ on the reference wave intensity $\tilde{I}_1^0$ for the case of thermal nonlinearity in the resonar absorbing medium:

( $\alpha_0 \ell$ = (1), 4 (2), and 8 (3); $\tilde{I}_3(0)/\tilde{I}_1(0) = 10^{-3}$ ).

small signal theory (see, for example, [12,13]) disregarding diffusive motion of the gas particles and Fig.10b illustrates the calculation results obtained on the basis of Eqs.(2.32). The curves 1–3 correspond to $\alpha_0 \ell = 2$, 4, and 8, respectively ($\ell$ is the length of the interaction region). We note that due regard for the diffusive spreading of amplitude gratings with small lifetimes essentially (from two-three times in the optimum up to one-two orders of magnitude at the ends of the curves) decreases the steady-state value of reflection coefficient in the media with resonant nonlinearity.[1]

Fig.11 shows the dependence of the reflection coefficient $R$ on the signal wave intensity at the fixed input intensities of the reference waves ($\tilde{I}_1(0)=\tilde{I}_2(\ell) = 10$, $\alpha_0 \ell = 8$). We see that the reflection coefficient is practically independent of the amplitude level of the signal wave up to the values $\tilde{I}_3(0)/\tilde{I}_1(0) \sim 10^{-1} \div 10^{0}$ (this corresponds to experimental data [7]) but it abruptly diminishes with further increasing $\tilde{I}_3(0)$ (the dashed line in Fig.11 gives the results of the small signal theory).

### 2.3.2. Stationary DFI in the media with thermal nonlinearity

It follows from the results of the previous section that the contribution of resonant nonlinearity to stationary DFI in absorbing media is usually small ($R \lesssim 10^{-2}$). Therefore, setting formally $\gamma_0 = 0$ in (2.31), we will now consider some features of DFI in the

---

[1] It was shown in [53] that the results of small signal theory [12] practically coincide with the results following from the equations of the given approach if the diffusive motion of gas particles is neglected.

case of thermal nonlinearity in the media with linear and nonlinear light absorption. At first, we neglect the variation of reference waves in the medium and restrict ourselves to small signal approximation

M e d i u m w i t h n o n l i n e a r a b s o r p t i o n. Suppose that $\alpha_1 = 0$ and $\alpha_0 \neq 0$. In the small signal approximation, we have $\Psi \simeq \tilde{C}_{1,1}/(1 + C_{0,0})$ and $\Psi/\Delta \simeq \tilde{C}_{1,1}/(1 + C_{0,0})^2$. Here $C_{0,0} = \tilde{I}_1 + \tilde{I}_2$, where $\tilde{I}_1 = b_1 b_1^*$ and $\tilde{I}_2 = b_2 b_2^*$ are the reference wave intensities.

As a result, the equations for signal and conjugated waves take the form

$$db_3/d\zeta = is_1(\tilde{I}_1 b_3 + b_1 b_2 b_1^*), \quad db_4^*/d\zeta = is_1(b_1^* b_2^* b_3 + \tilde{I}_2 b_4^*), \tag{2.33}$$

where $s_1 = \gamma_1 \alpha_0 Q/2(1 + \tilde{I}_1 + \tilde{I}_2)^2$.

Solving set (2.33) for the case of equal pump intensities ($\tilde{I}_1 = \tilde{I}_2$) and the boundary conditions of the form $b_3(0) = b_3^0$ and $b_4(l) = 0$, we obtain the following expression for the reflection coefficient:

$$R = \left|\frac{b_4(0)}{b_3(0)}\right|^2 = \mathrm{tg}^2\left(\frac{1}{2} s_1 \tilde{I} l\right), \tag{2.34}$$

where $\tilde{I} = \tilde{I}_1 + \tilde{I}_2$.

If $s_1 \tilde{I} l < 1$, then

$$R \cong \gamma_1^2 Q^2 (\alpha_0 l)^2 \tilde{I}^2 / 16(1 + \tilde{I})^4.$$

We see that in the latter case the extremum of $R$ is reached at $\tilde{I} = 1$. If $\tilde{I} \ll 1$, then $R \sim \tilde{I}^2$, but if $\tilde{I} \gg 1$, then $R \sim 1/\tilde{I}^2$.

M e d i u m w i t h l i n e a r l i g h t a b s o r p t i o n.

Now suppose that $\alpha_0 = 0$ and $\alpha_1 \neq 0$. Since $\rho = 0$, we obtain from (2.31)

$$db_3/d\zeta = is_2(\tilde{I}_1 b_3 + b_1 b_2 b_4^*), \quad db_4^*/d\zeta = is_2(b_1^* b_2^* b_3 + \tilde{I}_2 b_4^*),$$

where $s_2 = \gamma_1 \alpha_1 Q$.

If the intensities of the reference waves are equal, the reflection coefficient is determined in this case by relation (2.34) too (with the substitution $s_1 \to s_2$). However, $s_2$ does not depend on the radiation intensity now, thus for $s_2 \tilde{I} \ell < 1$ we obtain the well-known [6] result: $R \sim I^2$.

In the actual resonant media (with nonlinear light absorption), the intensities of reference waves vary appreciabely during their propagation, therefore a simple qualitative analysis does not give a a complete picture of DFI. Therefore it is useful to turn to the results of numerical integration of set of equations (2.31) in this situation.

First of all, we note that $(n_0 - 1) \sim 10^3$ in the gases at atmospheric pressure. Consequently, we have $\gamma_1 = k(n_0 - 1)/T \sim 10^{-2}$ in the visible and near-IR ranges at room temperature. Taking the typical values of the medium parameters $\rho \sim 10^{-3}$ g/cm$^3$, $c_V \sim 1$ J/(g x deg), $\sigma \sim 10^{-17}$ cm$^2$, $D \sim \varkappa \sim 0.1$ cm$^2$/s, and $\tau_1 \sim 10^{-6} \div 10^{-7}$ s, we have $\beta = h\nu/2\rho c_V \sigma \sim 1$. Thus we obtain $\gamma_1 Q \sim 10^0 \div 10^1$ for characteristic angles of convergence $\theta \sim 10^{-2}$ rad.

Fig. 12 illustrates the results of numerical integration of Eqs. (2.31) for the medium with nonlinear light absorption ($\alpha_1 = 0$, $\alpha_0 \neq 0$). The calculations were performed by an iteration method for $\gamma_1 Q = 10$. The exact resonance ($\gamma_0 = 1$) was considered and equal input intensities of the reference waves $I_1(0) = I_2(\ell)$ were taken.

It follows from Fig. 12 that the reflection coefficient of the signal wave in the media with small absorption of reference waves really corresponds to the values $\tilde{I}_1^0 = \tilde{I}_2^\ell \sim 1$. However, as $\alpha_0 \ell$ increases, the depletion of reference waves causes the displacement of the region of effective four-wave interaction towards the input, where the intensities of reference waves are higher. We note that there is an optimum value of $\alpha_0 \ell$ too, in particular, the extreme values of $R$ begin to diminish at sufficiently high small signal absorption (per pass).

## 2.4. DFI of multifrequency radiation in resonant media

Though a large number of references is devoted to investigation of degenerate four-wave interaction in resonant media, the features of DFI of multifrequency radiation that is generated by a set of gas lasers working on cascade transitions of two-atom molecules (for example, CO-lasers, lasers based on hydrohalides, etc.) were not studied in essence up to the present time. In this connection we can mention only the works [58-60] in one of which DFI of CO-laser radiation in liquid media with thermal nonlinearity was studied [58] and in the other two the attempts of theoretical analysis of the features of DFI of multifrequency radiation in multilevel gaseous resonant media were made. Below we develop an approach to the analysis of DFI of multifrequency radiation in the media with resonant nonlinearity on the basis of the results obtained in [60] for the simplest case of two-frequency laser radiation (it should be noted that these results can be generalized without any difficulty to the case of the media with thermal nonlinearity using the results of the previous section.

Restricting ourselves to an absorbing resonant medium, we consider DFI of counter-propagating waves in the case of two-frequency radiation on cascade vibrational transitions $0 \longrightarrow 1$ and $1 \longrightarrow 2$ of two-level molecules. For simplicity, the rotational structure of vibrational levels will be neglected.

We also assume that the profiles of spectral lines at the frequencies $\omega_1$ and $\omega_2$ do not overlap and that the summary $\omega_1 + \omega_2$ and difference $\omega_1 - \omega_2$ frequencies are far from the intrinsic characteristic frequencies of molecules.

Let us represent the total electromagnetic field $E^{(l)}$ at the frequency $\omega_l$ ($l$ = 1, 2) in the form of superposition of four running waves

$$E^{(l)} = \sum_{j=1}^{4} E_j^{(l)}, \text{ где } E_j^{(l)} = \mathscr{E}_j^{(l)} e^{i\omega_l t} + \text{c.c.} = A_j^{(l)} \exp\left[i\left(\omega_l t - \mathbf{k}_j^{(l)}\mathbf{r}\right)\right] + \text{c.c.}$$

Here, as before, $A_j^{(l)}(\mathbf{r}, t)$ is the slowly varying in time and space complex amplitude of the $j$th wave at the frequency $\omega_l$ and $\mathbf{k}_j^{(l)}$ is its wave vector.

Once again, we denote the field amplitudes of reference waves (directed in opposite directions) by $E_1^{(l)}$ and $E_2^{(l)}$, and the amplitudes of signal and conjugated waves by $E_3^{(l)}$ and $E_4^{(l)}$, respectively.

Representing the resonant polarization of the medium in the form of linear superposition of polarizations at the frequencies $\omega_1$ and $\omega_2$, we have $P = P^{(1)} + P^{(2)}$, where $P^{(l)} = \mathcal{P}^{(l)} \exp(i\omega_l t)$ + c.c. ($l$ = 1, 2). Restricting ourselves to the quasi-static approximation (with respect to the field) and considering homogeneous line broadening, let us use relations (1.18) for slowly varying in time complex amplitudes of polarization $\mathcal{P}^{(l)}$. Then, by analogy with (1.20), wave equation (1.3) gives

$$\sum_{j=1}^{4}\left(\mathbf{k}_j^{(l)} \frac{\partial A_j^{(l)}}{\partial \mathbf{r}} + \frac{1}{c} k^{(l)} \frac{\partial A_j^{(l)}}{\partial t}\right) e^{-i\mathbf{k}_j^{(l)}\mathbf{r}} = = -\frac{1}{2} k_l^{(l)} \gamma_l g_l \sum_{j=1}^{4} A_l^{(l)} e^{-i\mathbf{k}_j^{(l)}\mathbf{r}}, \quad l = 1, 2, \tag{2.35}$$

where $\gamma_\ell = 1 - i\delta_\ell$, $\delta_\ell$ is the frequency detuning and $g_\ell = \sigma_\ell(n_{\ell-1} - n_\ell)$ the frequency-dependent absorption coefficient at the frequency $\omega_\ell$, i.e. on the $\ell-1 \rightarrow \ell$ vibrational transition ($\sigma_\ell$ is the cross section for radiative absorption).

Using (1.16), we can represent the equations for absorption coefficients $g_1$ and $g_2$ in the form

$$\frac{\partial g_k}{\partial \tau} + \sum_{p=1}^{2} (f_p^{(k)} g_p + Q_p^{(k)} g_p \tilde{I}^{(k)}) = R_k + D\nabla^2 g_k, \quad k = 1,\ 2. \tag{2.36}$$

Here $\tilde{I}^{(k)} = I^{(k)}/I_s^{(k)}$ are the normalized radiation intensities at the frequencies $\omega_k$, $I_s^{(k)} = 1/2\,\sigma_k T_k$ are the saturation intensities for corresponding transitions, $I^{(k)} = \mathcal{E}^{(k)}\mathcal{E}^{(k)*}$, $\mathcal{E}^{(k)} = \sum_{j=1}^{4} \mathcal{E}_j^{(k)}$ and $\tau = t/T_1$ is the normalized time.

Supposing that the lower vibrational molecular levels (0, 1, and 2) of the resonant medium are adequately described by the harmonic oscillator model ($T_1/T_2 = 2$, $\sigma_2/\sigma_1 = 2$), we find the coefficients of Eqs. (2.36)

$$f_1^{(1)} = 0,\ f_2^{(1)} = -1,\ f_1^{(2)} = 2,\ f_2^{(2)} = 3,\ Q_1^{(1)} = 1,\ Q_2^{(1)} = -1/2,\ Q_1^{(2)} = -1,\ Q_2^{(2)} = 2,\ R_1 = 0 \text{ и } R_2 = 2\alpha_0,$$

where $\alpha_0 = \sigma_1 N_0$ and $N_0$ is the number of molecules per unit volume.

Introducing the notations (cf. with (2.11))

$$C_{0,0}^{(l)} = \sum_{j=1}^{4} A_j^{(l)} A_j^{(l)*}, \quad C_{2,0}^{(l)} = A_1^{(l)} A_2^{(l)*}, \quad C_{0,2}^{(l)} = A_3^{(l)} A_4^{(l)*},$$

$$C_{1,1}^{(l)} = A_1^{(l)} A_4^{(l)*} + A_2^{(l)*} A_3^{(l)}, \quad \tilde{C}_{1,1}^{(l)} = A_1^{(l)} A_3^{(l)*} + A_2^{(l)*} A_4^{(l)}, \tag{2.37}$$

where $A_k^{(\ell)}$ are the normalized (by $(I_s^{(k)})^{1/2}$) amplitudes of interacting waves, and taking into account the phase matching conditions for DFI of counter-propagating waves ($\mathbf{k}_1^{(\ell)} + \mathbf{k}_2^{(\ell)} = \mathbf{k}_3^{(\ell)} + \mathbf{k}_4^{(\ell)} = 0$), we obtain the expression for the total field intensities $\tilde{I}(l)$

$$\tilde{I}^{(l)} = C^{(l)}_{0,0} + (C^{(l)}_{2,0}\exp\{-i2\mathbf{k}^{(l)}_1\mathbf{r}\} + C^{(l)}_{0,2}\exp\{-i2\mathbf{k}^{(l)}_3\mathbf{r}\} + $$
$$+ C^{(l)}_{1,1}\exp\{-i\mathbf{k}^{(l)}_1\mathbf{r} - i\mathbf{k}^{(l)}\mathbf{r}\} + \tilde{C}^{(l)}_{1,1}\exp\{-i\mathbf{k}^{(l)}_1\mathbf{r} + i\mathbf{k}^{(l)}_3\mathbf{r}\} + (\text{c.c.}) \quad (2.38)$$

In view of (2.38) and (2.36), we seek the expressions for $g_l(\mathbf{r}, t)$ in the form of the following series expansions in spatial harmonics of the interacting waves at both frequencies:

$$g_l(\mathbf{r}, t) = g^{(l)}_{0,0} + \sum_{m=1}^{\infty}\sum_{s=0}^{2m}\{g^{(l)}_{2m-s,s}\exp[-i(2m-s)\mathbf{k}^{(1)}_1\mathbf{r} - is\mathbf{k}^{(1)}_3\mathbf{r}] + $$
$$+ \tilde{g}^{(l)}_{2m-s,s}\exp[-i(2m-s)\mathbf{k}^{(1)}_1\mathbf{r} + is\mathbf{k}^{(1)}_3\mathbf{r}] + \text{c.c.}\} + $$

$$+ \sum_{m=1}^{\infty}\sum_{s=0}^{2m}\{G^{(l)}_{2m-s,s}\exp[-i(2m-s)\mathbf{k}^{(2)}_1\mathbf{r} - is\mathbf{k}^{(2)}_3\mathbf{r}] + $$
$$+ \tilde{G}^{(l)}_{2m-s,s}\exp[-i(2m-s)\mathbf{k}^{(2)}_1\mathbf{r} + is\mathbf{k}^{(2)}_3\mathbf{r}] + \text{c.c.}\}, \quad l=1,2. \quad (2.39)$$

The coefficients $g^{(l)}_{p,s}$ in (2.39) correspond to the amplitude gratings induced at the $l-1 \rightarrow l$ transition by the interfering waves at the frequency $\omega_1$ and the coefficients $G^{(l)}_{p,s}$ correspond to the gratings induced at the same transition by the waves at the frequency $\omega_2$.

When $p, s \neq 0$, the quantities $C^{(l)}_{p,s}$, $g^{(l)}_{p,s}$ and $G^{(l)}_{p,s}$ in (2.37)-(2.39 are the symmetrized (with respect to the sign of the terms $p\mathbf{k}^{(l)}_1\mathbf{r}$ and $s\mathbf{k}^{(l)}_3\mathbf{r}$ in the exponents), while $\tilde{C}^{(l)}_{p,s}$, $\tilde{g}^{(l)}_{p,s}$, and $\tilde{G}^{(l)}_{p,s}$ are the antisymmetrized coefficients of the corresponding expansions in spatial harmonics. It is obvious that the coefficients $C^{(l)}_{0,0}$ and $g^{(l)}_{0,0}$ correspond to the constant components $\tilde{I}^{(l)}$ and $g_l$, while the coefficients $C^{(l)}_{2,0}$, $C^{(l)}_{0,2}$, $g^{(l)}_{2p,0}$, $g^{(l)}_{0,2p}$, $G^{(l)}_{2p,0}$, and $G^{(l)}_{0,2p}$ characterize the two-wave interaction of counter-propagating waves $E^{(l)}_1$, $E^{(l)}_2$ and $E^{(l)}_3$, $E^{(l)}_4$, respectively. Once again, let us take the $z$ axis along the diagonal between the directions of propagation of the waves $E_1^{(l)}$ and $E_3^{(l)}$ and restrict ourselves for simplicity to the case of small interaction angles ($\theta \ll 1$). Denoting the wave detuning between the fields at the frequencies $\omega_1$ and $\omega_2$ by $\Delta = k^{(1)} - k^{(2)}$ (we suppose that $\Delta \ll k^{(1)}, k^{(2)}$), we substitute (2.39) into (2.35) and collect the terms with identical rapidly oscillating factors of the form $\exp(\pm i\mathbf{k}_j^{(l)}\mathbf{r})$ in the left and right sides of the latter equation . If $\theta\Delta \ll 1$, then the slowly varying amplitudes $A_j^{(l)}$ are described by the equations of the form ($l = 1, 2$)

$$
\begin{aligned}
&\frac{\partial A_1^{(l)}}{\partial z} + \frac{1}{c}\frac{\partial A_1^{(l)}}{\partial t} = -\frac{1}{2}\gamma_l\,[Y_{0,0}^{(l)}A_1^{(l)} + e^{-i2(l-1)\Delta z}Y_{2,0}^{(l)}A_2^{(l)} + \\
&\qquad + \widetilde{Y}_{1,1}^{(l)}A_3^{(l)} + e^{-i2(l-1)\Delta z}Y_{1,1}^{(l)}A_4^{(l)}], \\
&-\frac{\partial A_2^{(l)}}{\partial z} + \frac{1}{c}\frac{\partial A_2^{(l)}}{\partial t} = -\frac{1}{2}\gamma_l\,[e^{i2(l-1)\Delta z}Y_{2,0}^{(l)*}A_1^{(l)} + Y_{0,0}^{(l)}A_2^{(l)} + \\
&\qquad + e^{i2(l-1)\Delta z}Y_{1,1}^{(l)*}A_3^{(l)} + \widetilde{Y}_{1,1}^{(l)*}A_4^{(l)}], \\
&\frac{\partial A_3^{(l)}}{\partial z} + \frac{1}{c}\frac{\partial A_3^{(l)}}{\partial t} = -\frac{1}{2}\gamma_l\,[Y_{1,1}^{(l)*}A_1^{(l)} + e^{-i2(l-1)\Delta z}Y_{1,1}^{(l)}A_2^{(l)} + \\
&\qquad + Y_{0,0}^{(l)}A_3^{(l)} + e^{-i2(l-1)\Delta z}Y_{0,2}^{(l)}A_4^{(l)}], \\
&-\frac{\partial A_4^{(l)}}{\partial z} + \frac{1}{c}\frac{\partial A_4^{(l)}}{\partial t} = -\frac{1}{2}\gamma_l\,[e^{i2(l-1)\Delta z}Y_{1,1}^{(l)*}A_1^{(l)} + \widetilde{Y}_{1,1}^{(l)}A_2^{(l)} + \\
&\qquad + e^{i2(l-1)\Delta z}Y_{0,2}^{(l)*}A_3^{(l)} + Y_{0,0}^{(l)}A_4^{(l)}].
\end{aligned}
\tag{2.40}
$$

Here we used the designations

$$
\begin{aligned}
&Y_{0,0}^{(l)} = g_{0,0}^{(l)}, \quad Y_{p,s}^{(l)} = g_{p,s}^{(l)} + G_{p,s}^{(l)}e^{i(p+s)\Delta z}, \\
&\widetilde{Y}_{p,s}^{(l)} = \widetilde{g}_{p,s}^{(l)} + \widetilde{G}_{p,s}^{(l)}e^{i(p-s)\Delta z}, \quad p,\ s = 1,\ 2,\ \ldots
\end{aligned}
\tag{2.41}
$$

The equations for coefficients of four-wave interaction are found from set of equations (2.36) taking account of (2.39) and using the relation between $k^{(1)}, k^{(2)}$ and $\Delta$ . As a result, an infinite (with respect to $p$ and $s$ ) chain of closed equations having a common structure is obtained for $Y^{(l)}_{p,s}$ and $\tilde{Y}^{(l)}_{p,s}$. For example, the equations for coefficients $Y^{(l)}_{1,1}$ and $\tilde{Y}^{(l)}_{1,1}$ can be written in the form

$$\begin{aligned}
&\frac{\partial Y^{(l)}_{1,1}}{\partial\tau} + \frac{1}{\tau^{(1)}_{1,1}} g^{(l)}_{1,1} + \frac{1}{\tau^{(2)}_{1,1}} G^{(l)}_{1,1} e^{i2\Delta z} + \sum_{p=1}^{2} f^{(l)}_{p} Y^{(p)}_{1,1} + \\
&\quad + \sum_{p=1}^{2} Q^{(l)}_{p} [C^{(p)}_{0,0} Y^{(p)}_{1,1} + e^{i(p-1)2\Delta z} (C^{(p)}_{2,0} \tilde{Y}^{(p)*}_{1,1} + C^{(p)}_{0,2} \tilde{Y}^{(p)}_{1,1} + \\
&\quad + C^{(p)}_{1,1} Y^{(p)}_{0,0}) + e^{-i(p-1)2\Delta z} (C^{(p)*}_{2,0} Y^{(p)}_{3,1} + C^{(p)*}_{0,2} Y^{(p)}_{1,3} + \\
&\qquad + C^{(p)*}_{1,1} Y^{(p)}_{2,2}) + \tilde{C}^{(p)}_{1,1} Y^{(p)}_{0,2} + \tilde{C}^{(p)*}_{1,1} Y^{(p)}_{2,0}] = 0, \\
&\frac{\partial \tilde{Y}^{(l)}_{1,1}}{\partial\tau} + \frac{1}{\tilde{\tau}^{(1)}_{1,1}} \tilde{g}^{(l)}_{1,1} + \frac{1}{\tilde{\tau}^{(2)}_{1,1}} \tilde{G}^{(l)}_{1,1} + \sum_{p=1}^{2} f^{(l)}_{p} \tilde{Y}^{(p)}_{1,1} + \\
&\quad + \sum_{p=1}^{2} Q^{(l)}_{p} [C^{(p)}_{0,0} \tilde{Y}^{(p)}_{1,1} + e^{i(p-1)2\Delta z} (C^{(p)}_{2,0} Y^{(p)*}_{1,1} + C^{(p)}_{0,2} \tilde{Y}^{(p)}_{1,3} + \\
&\quad + C^{(p)}_{1,1} Y^{(p)*}_{0,2}) + e^{-i(p-1)2\Delta z} (C^{(p)*}_{2,0} \tilde{Y}^{(p)}_{3,1} + C^{(p)*}_{0,2} Y^{(p)}_{1,1} + \\
&\quad + C^{(p)*}_{1,1} Y^{(p)}_{2,0}) + \tilde{C}^{(p)}_{1,1} Y^{(p)}_{0,0} + \tilde{C}^{(p)*}_{1,1} \tilde{Y}^{(p)}_{2,2}] = 0.
\end{aligned} \tag{2.42}$$

The constants $\tau^{(l)}_{1,1}$ and $\tilde{\tau}^{(l)}_{1,1}$ introduced in (2.42) are given by the relations $1/\tau^{(l)}_{1,1} = T_1/T^{(l)}_{1,1}$ and $1/\tilde{\tau}^{(l)}_{1,1} = T_1/\tilde{T}^{(l)}_{1,1}$, where

$$1/T^{(l)}_{p,s} = (k^{(l)})^2 D[(p+s)^2\cos^2\varphi + (p-s)^2\sin^2\varphi],$$

$$1/\tilde{T}^{(l)}_{p,s} = (k^{(l)})^2 D[(p-s)^2\cos^2\varphi + (p+s)^2\sin^2\varphi].$$

here $\varphi = \theta/2$ if $p, s \geq 1$ and $\varphi = 0$ if $p = 0$ or $s = 0$.

We do not give the equations for other coefficients $Y^{(l)}_{p,s}$ and $\tilde{Y}^{(l)}_{p,s}$ but in principle they can be written out without any difficulty. The first four members of such equations have the standard form, while the last one always contains nine terms including nine coefficients $C_{0,0}$, $C_{2,0}$, $C^{*}_{2,0}$, ..., $\tilde{C}^{*}_{1,1}$ (with the corresponding factors exp $[\pm i(p-1)2\Delta z]$) as one of the factors. And the second factor is given by different coefficients $Y_{p,s}$ and $\tilde{Y}_{p,s}$ or their complex conjugates. The concrete form of these factors is determined by the requirement that the correspon-ding index ($p$ or $s$) of the coefficient standing under the derivative sign should match the algebraic sum (with account of symmetrization and the sign of complex conjugation) of the same indices of the mentioned factors. As a result, the terms with identical exponential indices are separated out in (2.36).

The second and the third terms in (2.42) show that generally speaking the set of equations for the variables $Y^{(\ell)}_{p,s}$ and $\tilde{Y}^{(\ell)}_{p,s}$ is not closed if the particle diffusion is considered, because in the general case $T^{(1)}_{p,s} \neq T^{(2)}_{p,s}$. It is evident that mathematical meaning of this fact implies that in the case of multifrequency DFI in the presence of particle diffusion, representation of $g_\ell(r, t)$ in the form of a series in spatial harmonics of the interacting waves is actually impossible. And physically this is caused by the fact that diffusive motion of the particles results in mixing of the gratings formed by the fields at different frequencies. Therefore, the actual picture of interference interaction is much more complicated. However, in the two characteristic cases of widely separated or, on the contrary, very close frequencies $\omega_1$ and $\omega_2$, the set of equations for two-frequency DFI can be closed for either $g_{p,s}$ and $G_{p,s}$ or $Y_{p,s}$.

For example, in the case of close frequencies ($k^{(1)} \simeq k^{(2)} \simeq k$), the characteristic times of diffusive spreading are practically equal, so one may suppose that $T^{(\ell)}_{p,s} = T_{p,s}$ and $\tilde{T}^{(\ell)}_{p,s} = \tilde{T}_{p,s}$ for both frequencies ($\ell$ = 1, 2). Let us examine this case in more detail.

It as shown in section 2.1 that under the typical conditions of DFI, most of the coefficients $Y^{(\ell)}_{p,s}$ and $\tilde{Y}^{(\ell)}_{p,s}$ may be neglected because of the rather short times of diffusive spreading. Really, if the angle of wave divergence are small ($\theta \ll 1$), only the gratings corresponding to coefficients $\tilde{Y}^{(\ell)}_{p,p}$ have large characteristic times of diffusive spreading $\tilde{T}_{p,p} = 1/k^2 D p^2 \theta^2$ and consequently they appear to be long-lived. Therefore ,the DFI equations can be essentially simplified if we keep in them only those interaction coefficients that correspond to such gratings. As a result, the equations for the waves $A_j^{(\ell)}$ (disregarding the wave nonstationarity) take on the form ($\ell$ = 1, 2):

$$\begin{aligned}
\frac{\partial A_1^{(l)}}{\partial z} &= -\frac{1}{2}\gamma_l\,(Y_{0,0}^{(l)}A_1^{(l)} + \widetilde{Y}_{1,1}^{(l)}A_3^{(l)}),\\
-\frac{\partial A_2^{(l)}}{\partial z} &= -\frac{1}{2}\gamma_l\,(Y_{0,0}^{(l)}A_2^{(l)} + \widetilde{Y}_{1,1}^{(l)*}A_4^{(l)}),\\
\frac{\partial A_3^{(l)}}{\partial z} &= -\frac{1}{2}\gamma_l\,(Y_{0,0}^{(l)}A_3^{(l)} + \widetilde{Y}_{1,1}^{(l)*}A_1^{(l)}),\\
-\frac{\partial A_4^{(l)}}{\partial z} &= -\frac{1}{2}\gamma_l\,(Y_{0,0}^{(l)}A_4^{(l)} + \widetilde{Y}_{1,1}^{(l)}A_2^{(l)}).
\end{aligned} \tag{2.43}$$

The coefficients $Y_{0,0}^{(\ell)}$ and $\widetilde{Y}_{1,1}^{(\ell)}$ entering into (2.43) are determined by the infinite chain of bound equations

$$\begin{aligned}
&\frac{\partial Y_{0,0}^{(l)}}{\partial \tau} + \sum_{p=1}^{2}\,[(f_p^{(l)} + Q_p^{(l)}C_{0,0}^{(p)})\,Y_{0,0}^{(p)} +\\
&\qquad + Q_p^{(l)}\,(\widetilde{C}_{1,1}^{(p)}\widetilde{Y}_{1,1}^{(p)*} + \widetilde{C}_{1,1}^{(p)*}\widetilde{Y}_{1,1}^{(p)})] = R_l,\\
&\frac{\partial \widetilde{Y}_{m,m}^{(l)}}{\partial \tau} + \frac{1}{\tilde{\tau}_{m,m}}\,\widetilde{Y}_{m,m}^{(l)} + \sum_{p=1}^{2}\,[(f_p^{(l)} + Q_p^{(l)}C_{0,0}^{(p)})\,\widetilde{Y}_{m,m}^{(p)} +\\
&\qquad + Q_p^{(l)}\,(\widetilde{C}_{1,1}^{(p)}\widetilde{Y}_{m-1,m-1}^{(p)} + \widetilde{C}_{1,1}^{(p)*}\widetilde{Y}_{m+1,m+1}^{(p)})] = 0,\\
&\qquad m = 1,\ 2,\ 3,\ \ldots,
\end{aligned} \tag{2.44}$$

where $1/\tilde{\tau}_{m,m} = T_1 / \widetilde{T}_{m,m}$.

The analysis of set (2.43), (2.44) involves difficult calculation and cumbersome formulas even in the simpliest case of undepleted reference waves and small amplitudes of signal waves. Therefore it is convenient to turn to the results of numerical integration of these equations for finding out some features of two-frequency DFI. We note that since the modulus of coefficients $\widetilde{Y}_{m,m}^{(\ell)}$ decreases rapidly with increasing index $m$ of the grating, it is sufficient to use only some first equations of set (2.44) in the calculations. In particular, in the calculations described below, the coefficients with $m$ = 0, 1, ..., 5 were taken into account.

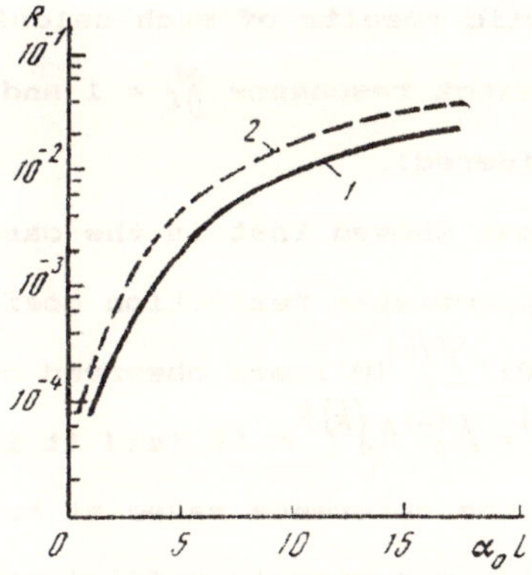


Fig. 13. Dependence of the reflection coefficient $R$ at the frequencie $\omega_1$ (1) and $\omega_2$(2) on the small-signal absoption $\alpha_0 \ell$ at the frequency $\omega_1$ for stationary two-wave DFI in the resonant medium: $\tilde{I}_1^{(k)}(0) = \tilde{I}_2^{(k)}(\ell) = 10$ and $\tilde{I}_3^{(k)}(0) = 10^{-2}$ ( $k$ = 1, 2).

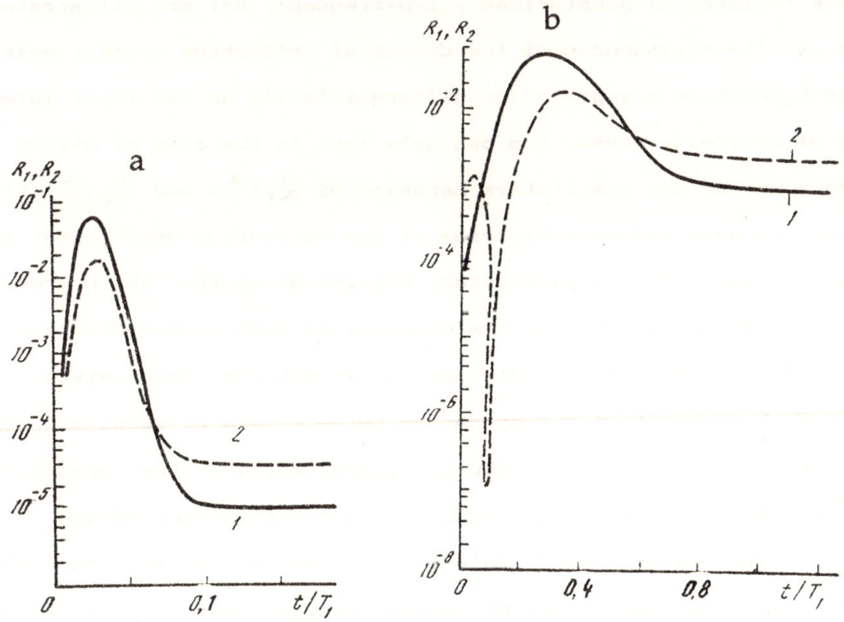


Fig. 14. Dynamical dependence of the reflection coefficients $R_1$ (1) and $R_2$ (2) on time at $\alpha_0 \ell$ = 4, $\tilde{I}_1^{(k)}(0) = \tilde{I}_2^{(k)}(\ell) = 10^2$ (a) and 10 (b), and $\tilde{I}_3^{(k)}(0) = 10^{-2}$ ( $k$ = 1, 2).

Some characteristic results of such calculations are illustrated by the figures (the exact resonance $\gamma_\ell = 1$ and small interaction angles $-\tilde{\tau}_{m_1 m} \gg 1$ were considered).

These calculations showed that in the case of stationary two-frequency DFI, appreciable reflection coefficients of the signal wave $R_k$ ( $R_k = \tilde{I}_4^{(k)}(0)/\tilde{I}_3^{(k)}(0)$ ) are observed only at the reference wave intensities $\tilde{I}_p^{(k)} = A_p^{(k)} A_p^{(k)*} \sim 10$ (and it is very important to have close intensities of the reference waves at both frequencies). If the small signal absorption per pass is sufficiently high ( $\alpha_0 \ell \gtrsim 10$, where $\ell$ is the length of the medium), then the reflection coefficients $R_1$ and $R_2$ exceed $10^{-2}$ (Fig. 13). Besides, the steady-state value of reflection coefficient at the second frequency exceeds the reflection coefficient in the $0 \longrightarrow 1$ band approximately by a factor of 1.5-2.

The features of nonstationary two-frequency DFI are illustrated in Fig. 14. The dependences of the dynamical reflection coefficients $R_1$ and $R_2$ on time are given for different levels of the input intensities of reference waves. One can note that in the case of strong reference waves, the qualitative behavior of $R_1(t)$ and $R_2(t)$ is analogous to nonstationary behavior of the reflection coefficient for single-frequency DFI. In particular, Fig. 14a shows that the extreme values of reflection coefficients obtained at both frequencies under essentially nonstationary conditions may exceed the steady-state values of these coefficients by one or three orders of magnitude. On the other hand, a rather deep dip can appear early in the dependence $R_2(t)$ (see Fig. 14b) if the level of input intensities of the reference waves is decreased. This is explained by the fact that the character of nonlinearity at the second frequency may vary in time as a result of a relatively slow population of the upper vibrational molecular levels (that are supposed to be initially unpopulated) under the action of the radiation field. In this case the sign of the interaction coefficient $\tilde{Y}_{1,1}^{(2)}$ appears to be different in various parts of the medium. Therefore the increase of the wave amplitude $A_4^{(2)}$ in one spatial region is compensated by its decrease in the other region. As a whole, this leads to an abrupt fall of the reflection coefficient for the 1 ⟶ 2 transition at a certain characteristic moment. We emphasize that such an effect is not observed under steady-state conditions.

In reality the features of DFI of multifrequency radiation in resonant media are not restricted to the mentioned phenomena. For example, the investigations of DFI in liquids performed with a cw CO laser [58], showed the beatings of the signal wave reflection coefficient under variation of the length of the interaction region. It is clear that such beatings (oscillations) result from spatial modulation of the medium parameters caused by the the interaction of the waves at close frequencies and they must be observed under the conditions when $\Delta\nu \ell \sim 1$ ($\Delta\nu$ is the difference between the frequencies $\omega_1$ and $\omega_2$ in $cm^{-1}$ and $\ell$ is the length of the medium). In principle, these features of DFI of multifrequency radiations directly follow from the analysis of the given general equations but we shall not dwell on this question because of rather cumbersome calculations.

## CONCLUSIONS

So, the general approach to the examination of both stationary and nonstationary multiwave interactions in resonant media at arbitrary field intensities and large number of spectral components of the radiation field is given in the quasi-static approximation with respect to the field. Interference (two-wave) interaction of multifrequency radiation in absorbing and amplifying media and degenerate four-wave interaction in the media with resonant and thermal nonlinearities are studied in detail.

It is shown that under certain conditions of two-wave (interference) interaction of multifrequency radiation in multilevel amplifying media, the asymmetry of amplification of counter-propagating waves may arise that actually results in single-directed amplification.

Some features of stationary and nonstationary DFI in resonant media are examined. In particular, it is shown that during the process of nonstationary DFI of two-frequency radiation (with close frequencies $\omega_1$ and $\omega_2$) in absorbing media on cascade transitions of two-atom molecules one can observe both an abrupt increase of the reflection coefficient of the signal wave at both frequencies and a deep dip at the frequency of the 1 $\longrightarrow$ 2 transition depending on the intensity level of the reference waves.

## REFERENCES


1. Kuznetsova T.I., Rautian S.G. "Plane solution of the wave equation for a layer with a negative coefficient of absorption with account of saturation" Izv. vuzov. Radiofizika., 1964, vol.7, No.4, p.682.
2. Ostrovskii L.A., Yakubovich E.I. "On the averaged laser equations and their stationary solutions" ZhETF, 1964, vol.46, No.3, pp.963-969.
3. Butylkin V.S., Kaplan A.E., Khronopulo Yu.G., Yakubovich E.I. "Rezonansnye vzaimodeistviya sveta s veshchestvom" [Resonance interactions of light with matter] Moscow: Nauka, 1977, 352 pp.
4. "Optical phase conjugation", Edited by R. Fisher, N.Y.: Acad.press, 1983.
5. Zel'dovich B.Ya., Pilipetskii N.F., Shkunov V.V. "Obrashcheniye volnovogo fronta" [Phase conjugation] Moscow: Nauka, 1985, 240 pp.
6. Bespalov V.I., Pasmanik G.A. "Nelineinaya optika i adaptivnye lazernye sistemy" [Nonlinear optics and adaptive laser systems] Moscow: Nauka, 1986, 136 pp.
7. Basov N.G., Kovalev V.I., Musaev M.A., Faizullov F.S. "Phase conjugation of a pulsed $CO_2$ -laser radiation" Trudy. FIAN, 1986, vol.172, pp.116-179.
8. Arutyunov Yu.A., Zherdienko V.V., Khizhnyak A.I. " Efficiency and quality of phase conjugation based on four-wave interaction in a co-propagation configuration" Izv. AN SSSR, Ser. fiz., 1987, vol.51, No.2, pp.347-357.
9. Lax M., Agrawal G.p., Belic M. et al. "Electromagnetic-field distribution in loaded unstable resonators" J. Opt. Soc. Amer. A, 1985, vol.2, No.5, pp.731-742.

10. Betin A.A., Milovskii N.D., Rul'kov N.F., Rusov N.Yu. "Propagation of spatially inhomogeneous radiation in a two-pass amplifier with a PC-mirror" Izv. vuzov. Radiofizika, 1986, vol.29, No.11, pp.1315-1323.

11. Abrams R.L., Lind R.C. "Degenerate four-wave mixing in absorbing media" Opt. Lett., 1978, vol.2, No.4, pp.94-96.

12. Gruneisen M.T., Gaeta A.L., Boyd R.W. "Exact theory of pump-wave propagation and its effect on degenerate four-wave mixing in saturable-absorbing media" J. Opt. Soc. Amer. B, 1985, vol.2, No.7, pp.1117-1121.

13. Gaeta A.l., Gruneisen M.T., Boyd R.W. Theory of degenerate four-wave mixing in saturable absorbing media with the inclusion of pump propagation effects" IEEE J. Quant. Electron., 1986, vol.22, No.7, pp.1095-1101.

14. Betin A.A., Dyatlov A.I., Kulagin S.N. et al. "Degenerate four-wave interaction in a resonant medium with the inclusion of variation of pump waves" Kvantovaya elektron., 1986, vol.13, No.10, pp.1975-1980.

15. Novikov A.D., Odulov S.G., Soskin M.S., Khizhnyak A.I. "Lasers based on degenerate four-wave interaction" Preprint of the Institite of Physics of AN of Ukraine, No.27, Kiev, 1986, 35 pp.

16. Heffernan D.M. "Transient degenerate four wave mixing in saturable absorbers" Phys. Lett. A, 1984, vol.103, No.5, pp.286-288.

17. Abrashin V.N., Apanasevich P.A., Afanas'ev A.A. et al. "Nonstationary phase conjugation based on four-wave interaction in resonant media" Kvantovaya elektron., 1985, vol.12, No.3, pp.546-552.

18. Apanasevich P.A., Afanas'ev A.A., Dritz V.V., Samson B.A. "Coherent forward four-wave mixing of ultrashort radiation pulses in a resonant medium" Opt. and Quant. Electron., 1986, vol.18, pp.129-135.

19. Lavoine J.P., Villaeys A.A. "Transient analysis of optical phase conjugation in absorbing media" Opt. Commun., 1986, vol.59, No.2, pp.160-166.

20. Cronin-Golomb M., Fisher B., White J.O., Yariv A. "A passive phase conjugate mirror based on self-induced oscillation in an optical ring cavity" Appl. Phys. Lett., 1983, vol.42, pp.919-921.

21. Cronin-Golomb M., Fisher B., White J.O., Yariv A. "Theory and application of four-wave mixing in photorefractive media" IEEE J. Quant. Electron., 1984, vol.20, No.1, pp.12-30.

22. Bel'dyugin I.M., Galushkin M.G., Zemskov E.M. "Phase conjugation of optical radiation using feedback based on four-wave interaction" Kvantovaya Elektron., 1984, vol.11, No.5, pp.887-893.

23. Odulov S.G., Slyusarenko S.S., Soskin M.S. "Energy characteristics of a laser with a phase conjugate mirror based on degenerate four-wave interaction in a medium with local response" Ibid., No.10, pp.2059-2068.

24. Betin A.A., Zhukov E.A., Mitropol'skii O.V. "On high reflection coefficients of four-wave mixing of $CO_2$ -laser radiation in liquids" Pis'ma v ZhTF, 1986, vol.12, No.7, pp.1052-1056.

25. Betin A.A., Sherstobitov V.E. " Methods and schemes for PC of radiation in the middle IR range" Izv. AN SSSR. Ser. fiz., 1987, vol.51, No.2, pp.299-306.

26. Bel'dyugin I.M., Galushkin M.G., Zolotarev M.V., Kamenets F.F. "Phase conjugation in a ring cavity with a four-wave mirror" Ibid., pp.358-361.

27. Betin A.A., Mitropol'skii O.V. "Generation of radiation using four-wave interaction in a scheme with feedback in the range $\lambda$ = 10 μm" Kvantovaya Elektron., 1987, vol.14, No.5, pp.1002-1008.

28. Betin A.A., Rusov N.Yu. "On the conjugated wave buildup in a scheme of four-wave interaction using feedback" Izv. vuzov. Radiofizika, 1987, vol.30, No.5, pp.676–678.

29. Erofeenko V.T., Urbanovich A.I. "Deduction of equations for optical phase conjugation based on four-wave interaction in in resonant media in the "following" nonlinear polarization mode" Moscow, 1987, 21 p., Dep. v VINITI AN SSSR 14.11.1987, No.4566-B 87.

30. Letokhov V.S., Chebotaev V.P. "Printsipy nelineinoi lazernoi spektroskopii" [Principles of nonlinear laser spectroscopy] Moscow: Nauka, 1975, 280 pp.

31. Pantell R., Puthoff H. "Osnovy kvantovoi elektroniki" [Fundamentals of Quantum Electronics] Moscow: Mir, 1972, 384 pp.

32. Blombergen N. "Nelineinaya optika" [Nonlinear optics] Moscow: Mir, 1966, 424 pp.

33. Bel'dyugin I.M., Stepanov A.A., Shcheglov V.A. "On the deduction of equations of degenerate two-wave interaction with the inclusion of nonstationary response of a resonant medium" Kratk. soobshch. po fizike FIAN, 1986, No.9, pp.40–42.

34. Virnik Ya.Z., Piskunov A.K., Stepanov A.A., Shcheglov V.A. "Diffraction effects in cw chemical lasers with an unstable confocal resonator" Kvantovaya elektron., 1979, vol.6, No.1, pp.236–248.

35. Bel'dyugin I.M., Stepanov A.A., Shcheglov V.A. "On the theory of two-wave (interference) interaction of multifrequency radiation on cascade transitions in multilevel resonant media" Ibid., 1988, vol.15, No.3, pp.531–538.

36. Gordiets B.F., Osipov A.I., Shelepin L.A. "Kineticheskie protsessy v gazakh i molekulyarnye lazery" [Kinetic processes in gases and molecular lasers] Moscow: Nauka, 1980, 512 pp.

37. Bel'dyugin I.M., Stepanov A.A., Shcheglov V.A. "On the theory of degenerate two-wave (interference) interaction of two-frequency radiation in resonant media" Preprint FIAN, No.195, Moscow, 1987, 28 pp.

38. Korn G., Korn T. "Spravochnik po matematike dlya nauchnykh rabotnikov i inzhenerov" [Mathematical handbook] Moscow: Nauka, Nauka, 1973, 832 pp.

39. Hoffman H.J. "Thermally induced phase conjugation by transient real-time holography: A review" J. Opt. Soc. Amer. B, 1986, vol.3, No.2, pp.253-273.

40. Steel D.G., Lind R.C., Lam J.F. "Degenerate four-wave mixing at resonant homogeneously broadened system" Phys. Rev. A, 1981, vol.23, No.5, pp.2513-2524.

41. Lind R.C., Steel D.G., Dunning G.J. "Phase conjugation by resonantly enhanced degenerate four-wave mixing" Opt. Eng., 1982, vol.21, No.2, pp.190-198.

42. Vasil'ev L.A., Galushkin M.G., Seregin A.M., Cheburkin N.V. "Theoretical investigation of the phase conjugation efficiency in inverted carbon dioxide" Izv. AN SSSR. Ser. fiz., 1981, vol.45, No.8, pp.1412-1416.

43. Vasil'ev L.A., Galushkin M.G., Seregin A.M., Cheburkin N.V. "Phase conjugation in inverted carbon dioxide caused by light-induced thermal deposition" Kvantovaya elektron., 1982, vol.9, No.6, pp.1228-1233.

44. Vasil'ev L.A., Galushkin M.G., Seregin A.M., Cheburkin N.V. "Phase conjugation based on four-wave interaction in a medium with thermal nonlinearity" Ibid., No.8, pp.1571-1575.

45. Caro R.G., Gower M.C. "Phase conjugation by degenerate four-wave mixing in absorbing media" IEEE J. Quant. Electron., 1982, vol.18, No.9, pp.1376-1380.

46. Brown W.P. "Absorption and depletion effects on degenerate four-wave mixing in homogeneously broadened absorbers" J. Opt. Soc. Amer., 1983, vol.73, No.5, pp.629-634.

47. Galushkin M.G., Seregin A.M., Fedorov A.B., Cheburkin N.V. "Effect of excited particle diffusion on phase conjugation in a gas medium" Kvantovaya Elektron., 1983, vol.10, No.10, pp.2115-2118.

48. Novikov A.D., Odulov S.G., Soskin M.S., Khizhnyak A.I. "Lasers based on degenerate four-wave interaction" Preprint of the Institute of Physics of AN of Ukraine, No.27, Kiev, 1986.

49. Onoshko R.N., Rubanov A.S., Tolstik A.L. "Effect of diffusive erasing of small-scale gratings on the efficiency of phase conjugation in a resonant medium" ZhPS, 1986, vol.45, No.3, pp.315-318.

50. Goryachkin D.A., Kalinin V.P., Kozlovskaya I.M. et al. "Investigation of degenerate four-wave interaction in $SF_6$ " Optika i spektroskopiya, 1986, vol.60, No.2, pp.324-330.

51. Galushkin M.G., Zemskov E.M., Klushin V.N. et al. "Degenerate four-wave interaction in a medium with light-induced thermal deposition at chemical reactions" Izv. AN SSSR. Ser. fiz., 1987, vol.51, No.2, pp.311-317.

52. Galushkin M.G., Sviridov K.A., Seregin A.M., Cheburkin N.V. "Degenerate four-wave interaction in a medium with a thermal chemical nonlinearity" Ibid., pp.318-322.

53. Bel'dyugin I.M., Zolotarev M.V., Stepanov A.A., Shcheglov V.A. "On the theory of nonstationary DFI in resonant media in the counter-propagation configuration at arbitrary field intensities" Kvantovaya Elektron., 1987, vol.14, No.11, pp.2317-2324.

54. Bel'dyugin I.M., Lepeshkin A.A., Stepanov A.A., Shcheglov V.A. "Theory of resonant degeneration of four-wave interaction (DFI) at arbitrary field intensities with the inclusion of particle diffusion in a gas" Optika i spektroskopiya, 1987, vol.63, No.4, pp.1207-1212.

55. Bel'dyugin I.M., Stepanov A.A., Shcheglov V.A. "On the theory of DFI in the counter-propagation configuration in the media with resonant and thermal nonlinearity mechanisms at arbitrary field intensities" Kvantovaya elektronika, 1989, vol.16, No.1, pp.84-90.

56. Basov N.G., Kovalev V.I., Musaev M.A., Faizullov F.S. "Investigation of reflection for four-wave interaction in resonant gases at 10.6 μm" Preprint of the Lebedev Physics Institute, No.204, Moscow, 1981, 16 pp.

57. Basov N.G., Kovalev V.I., Lesov A.R., Faizullov F.S. "Investigation of phase conjugation of a pulsed $CO_2$ -lase radiation using four-wave interaction in $SF_6$ " Pis'ma v ZhTF, 1982, vol.8, No.8, pp.451-455.

58. Betin A.A., Zhukov E.A., Novikov V.P. "Four-wave interaction of a CO-laser radiation in carbon tetrachloride" Optika i spektroskopiya, 1985, vol.59, No.6, pp.1363-1366.

59. Galushkin M.G., Nikitin V.Yu., Oraevskii A.N. "PC of an optical wave using degenerate four-wave interaction in an amplifying medium of chemical HF-lasers" Kvantovaya elektronika, 1988, vol.15, No.1, pp.153-160.

60. Bel'dyugin I.M., Stepanov A.A., Shcheglov V.A. "On DFI of multifrequency radiation in resonant media at cascade transitions of two-level molecules" Ibid., No.7, pp.1480-1487.

# EFFECT OF NONMONOCHROMATICITY OF OPTICAL BEAMS ON THE EFFICIENCY AND QUALITY OF FOUR-WAVE INTERACTION

*A.I. Erokhin, A.A. Stepanov and V.A. Shcheglov*

## 1. ON THE FEATURES OF PHASE CONJUGATION BASED ON FOUR-WAVE INTERACTION

The questions associated with investigation of the effect of nonmonochromaticity of optical beams on the efficiency and quality of four-wave interaction (FWI) have a direct relatonship to the problem of phase conjugation.

The investigations of phase conjugation (PC) of laser radiation in nonlinear media received wide acceptance after the discovery of the phenomenon of self-conjugation of a wave front in the process of stimulated Brillouin scattering (SBS) [1]. Another promising method of phase conjugation is based on FWI. In fact, this metod is proposed in holography [2-4], while in real time it was realized in [5,6]. Later on the PC method was proposed that successfully combines the advantages of the above-mentioned techniques [7].

PC based on FWI has some advantages over the self-conjugation method: zero threshold, fast response, simplicity of interpretation, possibility of simultaneous phase conjugation and signal amplification, and also the possibility of PC in those spectral ranges where SBS has not been reliably realized yet. We note that at present FWI is the only PC method in the middle-IR range. PC of a $CO_2$-laser radiation was first observed in [8]. The chief disadvantage of the FWI method is the contradictory requirement of forming high-power conjugate reference waves that are coherent with the radiated signal wave being conjugated. However, this disadvantage has no effect on the applications of this method for investigation of physical properties of matter.

Consider PC based on FWI. This method consists in the following. Two counter-propagating waves $E_1$ and $E_2$ (reference waves) interact with a third (signal) wave $E_3$ in a medium having the third order nonlinear polarizability $\chi^{(3)}$. Generally speaking, the signal wave has an arbitrary wave front and may be written in the form

$$\mathbf{E}_3(\mathbf{r}, t) = {}^1\!/_2\mathbf{A}_3(\mathbf{r}) \exp(i\omega t) + \text{c.c.} \tag{1}$$

Wave interaction results in generation of the fourth wave $E_4$. Its wave front is conjugated with respect to the wave front of the signal wave and it propagates directly in the opposite direction:

$$\mathbf{E}_4(\mathbf{r}, t) = {}^1\!/_2\mathbf{A}_3^*(\mathbf{r}) \exp(i\omega t) + \text{c.c.} \tag{2}$$

At present FWI is used for conjugation of high-power optical beams in combination with PC based on stimulated scattering [7,9].

The main physical properties of FWI are much studied in the degenerate case, i.e. in the case when the interacting monochromatic waves have equal frequencies [10-13].

In the degenerate case, when the interacting waves are monochromatic and their frequencies are equal, the appearance of the wave $E_4$ can be explained in the terms of holography [14,15]. When the co-propagating waves $E_1$ and $E_3$ interfere, a stationary holographic grating of the dielectric permittivity $\delta\varepsilon_{13} \sim \chi_{13}^{(3)} A_1 A_3^*$ is recorded in the medium. It has the period $\Lambda_{13} = (\lambda/2)\sin(\theta/2)$, where $\lambda$ is the radiation wavelength and $\theta$ is the convergence angle of the waves $E_1$ and $E_3$ (Fig. 1a). The counter-propagating waves $E_2$ and $E_3$ record a transverse grating $\delta_{23} \sim \chi_{23}^{(3)} A_2 A_3^*$ in the medium with the period $\Lambda_{23} \simeq \lambda/2$ (Fig. 1b)[1].

The longitudinal holographic grating $\delta_{13}$ is reconstructed by the wave $E_2$, while the transverse grating $\delta_{23}$ is reconstructed by the wave $E_1$. In each case reconstruction results in the appearance of the fourth (reflected) wave $E_4$. As a matter of fact, $E_4$ is the combined effect of two reconstructions. The direction of propagation of the reflected wave $E_4$ is determined by the Bragg condition for volume gratings:

$$\mathbf{k}_4 = \mathbf{k}_1 - \mathbf{k}_3 + \mathbf{k}_2, \qquad (3)$$

where $\mathbf{k}_i$ are the wave vectors of the corresponding waves. In stationary holography, the frequency of the wave $E_4$ is equal to the frequency of the reconstructing wave. Therefore, $\mathbf{k}_4 = -\mathbf{k}_3$ in the degenerate case, i.e. the direction of propagation of the reflected wave is exactly opposite to the direction of the signal wave.

---

[1] If the nonlinear suspectibility $\chi^{(3)}$ instantaneously follows the field, then the nonstationary term appears, which is associated with modulation $\delta\varepsilon$ at the double frequency $2\omega$. This term is caused by interference of counter-propagating pump waves $E_1$ and $E_2$ (parametric mechanism of PC). In the case of liquids characterized by inertia of nonlinear properties, this modulation has little effect on the reflected signal.

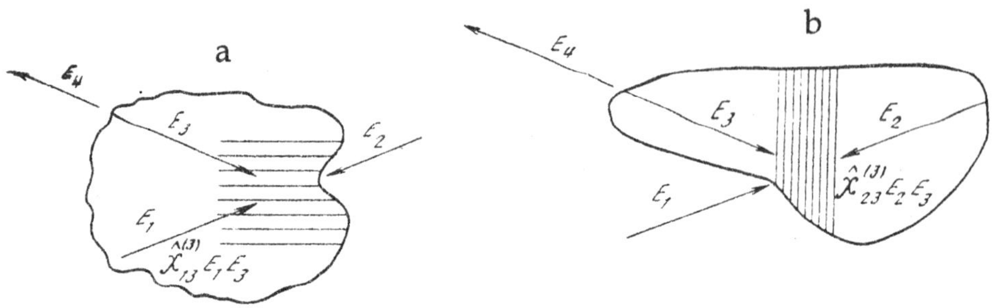


Fig. 1. Recording and reconstruction of longitudinal (a) and transverse (b) gratings of the refractive index perturbation using PC based on FWI.

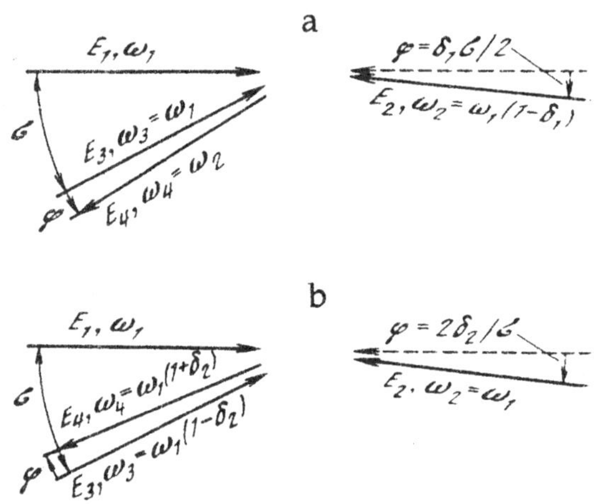


Fig. 2. Nondegenerate FWI under phase matching conditions. Degeneration is removed in the frequency of the reference wave counter-propagating with respect to the signal (a), in the frequency of the signal wave (b

Mathematical description of FWI identical to the holographic one is usually given in the terms of nonlinear polarization [16]:

$$P_{NL} = {}^1\!/_2\chi^{(3)} A_1 A_2 A_3^* \exp(i\omega t) + \text{c.c.}$$

In some cases of practical importance (with regard to the above-mentioned remarks) it is advantageous to divide the nonlinear polarization $P_{NL}$ into the terms:

$$P_{NL} \infty A_2 (\chi_{13}^{(3)} A_1 A_3^*) + A_1 (\chi_{23}^{(3)} A_2 A_3^*) + \text{c.c.} \qquad (4)$$

This proves itself, for example, for different polarizations of interacting waves, in the case of nonlocal nonlinear properties of a medium [17], and also in the case when the inertia of nonlinear suspectibility $\chi$ is taken into account.

## 2. NONDEGENERATE FOUR-WAVE INTERACTION

### 2.1. Nondegenerate four-wave interaction under phase matching conditions

In contrast to the degenerate case (equal frequencies of all the fourWaves), PC based on nondegenerate FWI has not been adequately studied. Nondegenerate FWI differs from its degenerate analogue both in the reflection efficiency and the quality of PC. This may become apparent even in the case when the frequency of only one of the interacting waves differs from the two others. This difference may be some orders of magnitude smaller than the frequency itself.

The necessity of considering PC based on FWI in a nondegenerate case [12.13,18-23] is caused by the practical requirements. First, the signal wave frequency may be diffrent from the frequencies of all the reference waves. Second, the spectrum of most of high-power lasers has a linear structure and in principle cannot be represented in the form of single-frequency radiation. It should be noted that frequency detuning of one of the waves allows to carry out investigation on active spectroscopy of nonlinear media used for PC. In the case under consideration, the FWI process becomes essentially complicated at least by the fact that the standing gratings formed by interference of the waves with equal frequencies transform into the running ones when the frequencies are different. This change from the standing holograms to the running ones may cause the change of both the direction of reflection and the reflection efficiency.

In the nondegenerate case, the efficiency of hologram reconstruction decreases as a result of breaking the phase matching condition (3) since the waves reflected by different points of a medium add up out of phase. According to [21,24], the reflection coefficient (the ratio of reconstructed and signal wave intensities) essentially depends on the wave vector mismatch $\Delta k = |\mathbf{k}_1 + \mathbf{k}_2 - \mathbf{k}_3 - \mathbf{k}_4|$:

$$R = I_4/I_3 \infty \left(\frac{\sin \Delta kl/2}{\Delta kl/2}\right)^2, \qquad (5)$$

where $l$ is the length of the region of nonlinear interaction. The phase matching criterion is obvious: $\Delta k l \ll 1$. Phase matching is not obligatory for the achievment of high-efficiency FWI in those situations when the FWI increment (defined in [16] as $\varkappa = (2\pi\omega/cn)\chi^{(3)} A_1 A_2$ exceeds $\Delta k$, i.e. $\varkappa > \Delta k$ (the case of nonlinear matching).

Some theoretical and experimental works [23,25-29] are devoted to the methods of obtaining matching condition in nondegenerate FWI (the case $\ae < \Delta k$). As a matter of fact, realization of these methods reduces either to turning the direction of propagation or to changing the frequency of one of the reference waves. For example, if the frequency detuning of one of the reference waves is fixed [23,25] ($\omega_1 = \omega_3 \neq \omega_2$), it is proposed to turn the frequency-detuned reference wave by an angle $\varphi$ towards the signal wave. For small $\sigma$ (the divergence angle between the signal and the reference waves with equal frequencies), $\varphi$ is determined by the expression $\varphi \simeq \delta_1 \sigma / 2$, where $\delta_1 = (\omega_1 - \omega_2)/\omega_2$. In this case the reflected radiation deflects from the direction of the $-k_3$ vector by an angle $-\varphi$ (Fig.2a). If degeneration is removed in the frequency of the reference wave ($\omega_1 = \omega_2 \neq \omega_3$), then phase matching is achieved by turning the reference wave (counter-propagating with respect to the signal wave) by an angle $\varphi \simeq 2\delta_2/\sigma$ [16-18], where $\delta_2 = (\omega_1 - \omega_3)/\omega_1$. In this case, reflected radiation deflects from the direction of the $-k_3$ vector by an angle $\varphi$ (Fig.2b).

Another method of achieving phase matching for the frequency-detuned signal wave is proposed in [28,29]. The frequency of one of the reference waves is appropriately tuned to have equal frequencies of all the interacting waves in one of the moving inertial systems.

## 2.2. Nondegenerate four-wave interaction without phase matching

The question of the direction of propagation of a "conoscopic" wave (the case of removing degeneration in the frequency of the signal wave is meant) was first raised by Yu.N. Denisyuk [30]. He pointed out that in the general case the direction of propagation of the reflected wave differs from the one of the conjugated wave.

An analogous question was raised in the theoretical work [31] devoted to PC based on FWI. The author obtained an integral representation for the reflected wave in an approximation of a strongly bounded volume of nonlinear interaction and showed on its basis that an image of an object increases its dimensions when degeneration is removed. In this case the coefficient of magnification is equal to the ratio of the frequencies of reflected and signal waves.

The expression (5) describing the effect of phase mismatch on the efficiency of FWI was obtained in [21] as a result of solving a problem, which was actually one-dimensional. It follows from (5) that the factor of FWI efficiency decrease is essential at $\Delta k l \gg 1$. Apparently, this circumstance brought the authors of [21] to the conclusion that under the conditions of phase mismatch the direction of the reflected wave corresponds to the minimum $\Delta k l$ value and consequently to the minimum $\Delta k$, but this is not quite correct. The principle of minimum $\Delta k$ was also used in [19,23] for determination of the direction of reflected radiation in the absence of phase matching. In particular, in the case of removing degeneration in one of the reference waves ($\omega_1 = \omega_3 \neq \omega_2$), the expression for the angle of deflection of the reflected wave with respect to the ideally conjugated one was obtained:

$$\theta \approx \text{arc tg}\left(\frac{\text{tg}\,\sigma}{1+\delta_1\cos\sigma}\right) - \sigma. \qquad (6)$$

When $\delta_1$ and $\sigma$ are small, this formula corresponds to the case, which is close to phase matching. The direction of reflection in the case $\omega_1 = \omega_2 \neq \omega_3$ under the conditions of strong nonlinear interaction ($\varkappa \gg \Delta k$ and $R > 1$) was experimentally determined in [33]. It should be noted that in principle the result obtained in [33] may differ from the data for the case $\varkappa \ll \Delta k$ .

Theoretical work [27] devoted to nondegenerate FWI took into consideration the presence of a boundary of a nonlinear medium. It gave the general equation with the inclusion of this boundary, however its effect on the properties of reflected radiation was not analyzed there.

The effect of boundaries of the region of nonlinear interaction and their orientation on the efficiency and the direction of a conjugated wave reflection in the case of nondegenerate FWI was investigated in [34] both theoretically and experimentally. In particular, it was found that at small dimensions of the interaction region $\ell$ the reflection efficiency increased according to the quadratic law $(\sim \ell^2)$, as in the degenerate case, but then oscillations with a constant amplitude appeared. The oscillatory character of this dependence was attributed to the presence of boundaries and was caused by interference of the fields reflected from the boundaries of a nonlinear medium. It was found that in the case of a single boundary and the absence of spatial phase matching, FWI gave reflection only at the subboundary layer of a nonlinear medium. In the particular case of removing degeneration only in the signal wave frequency, the direction of reflection depended only on the angle of incidence of the signal wave on the boundary of a nonlinear reflector.

## 3. FOUR-WAVE INTERACTION OF MULTIFREQUENCY WAVES

In some cases of practical importance, one meets a situation when each of the four interfering signals participating in FWI contains several spectral components. One of the first works devoted to FWI of multifrequency waves are [18-20]. A DF-laser with a spectrum containing a number of rotational-vibrational lines was used in experiments [18], while a $CO_2$-laser oscillating on several P- and R-lines was taken in [19,20] for the study of FWI. In the cited works, radiation of one and the same laser was used for forming the signal and the reference waves.

DF-laser radiation consisting of separate spectral lines with a widths less than 1 $cm^{-1}$ (in a spectral range 3.6-4.0 μm) was used in [18]. A germanium plate 5 mm thick (refractive index $n \simeq 4$) was taken as a nonlinear medium. The signal and the reference waves converged at an angle of $\sim 0.1$ rad. It was found that the conversion efficiency diminished by a factor of nine when single-frequency oscillation was changed by multifrequency oscillation on several lines (five of them were the most intensive). If the whole process is considered as the superposition of independent degenerate FWI processes on each separate line, then within the limits of this approach one should expect the decrease of the interaction efficiency by a factor of 25 for the change from one to five spectral lines. We note that the question of matching the signal and the reference waves was not considered in [18].

A $CO_2$-laser oscillating simultaneously on two lines P(16) and P(14) ($\Delta\nu \simeq 1.7$ $cm^{-1}$) was used in [19]. The nonlinear medium was a gas cell 2 cm long filled with $SF_6$. The signal and reference waves converged at an angle $\sim 4 \cdot 10^{-2}$ rad. This minimum angle was specially chosen to eliminate the possible disturbance of phase matching conditions. Reflection was observed on the same spectral lines as the laser radiation that is indicative of the absence of cross-interactions.

The possibility of PC of a $CO_2$-laser radiation with a spectrum consisting of several P- and R-lines was demonstrated in [20]. The maximum laser bandwidth was $\Delta\nu \simeq 20$ $cm^{-1}$. The signal wave and the co-propagating reference wave were thoroughly matched in a nonlinear medium, a Ge sample 1 cm thick. They converged at an angle $\delta \sim 6 \cdot 10^{-2}$ rad. It was experimentally determined that the reflection coefficient

$R$ decreased by a factor of 1.3 for the change from single-frequency oscillation to multifrequency oscillation on several lines within a bandwidth $\Delta\nu \simeq 20\ cm^{-1}$). The theory of multifrequency FWI for collinear inertialess interaction is developed in [20]. In particular, it is shown that in the case of short nonlinear media ($\ell \ll 2\pi c/n\Delta\nu$) the efficiency of FWI becomes several times higher as one passes from monochromatic to multifrequency waves, while it remains constant in the case of long media.

## 4. THEORY OF FOUR-WAVE INTERACTION OF NONMONOCHROMATIC RADIATION IN THE MEDIA WITH RELAXATION

### 4.1. General analysis (Born approximation)

Laser radiation has one or another degree of nonmonochromaticity depending on the actual conditions. So, the relative line width $\Delta\omega/\omega$ of high-power Q-switched lasers lies in the range from $10^{-8}$ to $10^{-2}$. Therefore, the study of PC of broadband radiation that is characteristic of high-power lasers (in particular, solid-state lasers), is of certain interest for practical applications. Large line width, multimode oscillation regime, and, as a consequence, insufficient temporal coherence of radiation of such systems result in the principal features of FWI in this case.

Although the experiments on four-wave interaction of multifrequency radiation [18-20] showed the possibilities of PC of multifrequency signals, nevertheless they did not give unumbiguous conclusions. So, in [18] the reflection coefficient considerably decreased with increasing the spectral width, while in [20] it remained practically equal to the reflection coefficient of a monochromatic signal. The principle possibility of PC in a multifrequency case was shown in [19], however the analysis of the results obtained and of the PC quality was not carried out.

We note that a theoretical analysis of the efficiency of PC of nonmonochromatic radiation for the case of inertialess nonlinearity and small convergence angles (collinear geometry) was performed in [20]. The same analysis was extended in [35] to the case of nonlinear media with relaxation. In contrast to [20] it was assumed that all the three initial waves were obtained from different sources. The data given below is guided by the results of [35].

Let us consider four-wave interaction of broadband radiation in a nonlinear media with relaxation. We denote the field amplitudes of counter-propagating reference waves by $E_1$ and $E_2$ and the amplitudes of signal and conjugated waves by $E_3$ and $E_4$, respectively. For simplicity, we suppose an angle between the directions of propagation of the waves $E_1$ and $E_3$ to be small. Restricting ourselves later on to the Born approximation, let us also suppose that the amplitude of the conjugated wave $E_4$ is small compared with the others. Then one may neglect the reciprocal action of the wave $E_4$ on the waves $E_1$, $E_2$, and $E_3$ (depletion of the waves in a nonlinear medium is not taken into account).

Let $\Delta\omega = 2\pi\Delta\nu$ be the frequency band of the interacting waves (radiation bandwidth) and $\omega$ be the radiation line center frequency. Supposing that $\Delta\omega / \omega \ll 1$, nevertheless we shall consider a large number of independent modes (for example, axial modes of the cavity), that are actually unlocked, within the radiation bandwidth.

It is clear that under these conditions the reflected signal intensity is a random function varying with the same time scale ($\Delta t \sim 1/\Delta\nu$) as the amplitudes of the interacting waves. The commonly used radiation detectors have a rather slow response ($T_d \gg 1/\Delta\nu$) and do not respond to fast radiation fluctuations (integrating them with a time constant $T_d$ ). Therefore, later on we shall care only for the values of the reflected wave intensity obtained by averaging $I_4$ in a time interval, which is considerably greater than the characteristic scale of temporal correlation of the fields. An averaging procedure is discussed in detail in [36] when solving the problem of self-diffraction of nonmonochromatic fields.

In the general formulation, the problem is reduced to solving the wave equation

$$\Delta E_4 - \left(\frac{n}{c}\right)^2 \frac{\partial^2}{\partial t^2} E_4 = \left(\frac{4\pi}{c^2}\right) \frac{\partial^2}{\partial t^2} P_{NL} \tag{7}$$

( $P_{NL}$ is the nonlinear polarization of a medium) with the following averaging of the quantity $I_4 = E_4 E_4^*$.

If the waves $E_1$ and $E_3$ propagate along the $x$ axis, while the waves $E_2$ and $E_4$ have the opposite direction, then it is evident that the wave equation for space-time dependences of the waves $E_{13}(x,t)$ and $E_2(x,t)$ gives in the Born approximation

$$E_{1,3}(x, t) = E_{1,3}(t - x/c), \qquad E_2 = E_2(t + x/c), \tag{8}$$

where $c$ is the velocity of propagation of the waves.

In the same approximation, for the case of a relatively short nonlinear medium, we may write the following expression for the conjugated wave amplitude at a sufficiently large distance from the interaction region [37]

$$E_4(x, t) = A \int_0^l P_{NL}(x', t - |x - x'|/c)\, dx', \tag{9}$$

where $A$ is some constant (integral (9) is taken along all the length $l$ of the nonlinear medium).

Then, considering a medium with relaxation and neglecting electron nonlinearity (see, for example, [15]), one can represent the polarization responsible for backward reflection of a signal in the form [38, 39]

$$P_{NL}(t) = \frac{1}{\tau_{13}} \chi_{13} E_2(t) \int_{-\infty}^{t} E_1(\xi) E_3^*(\xi) \exp\left[-\frac{t-\xi}{\tau_{13}}\right] d\xi + \\ + \frac{1}{\tau_{23}} \chi_{23} E_1(t) \int_{-\infty}^{t} E_2(\xi) E_3^*(\xi) \exp\left[-\frac{t-\xi}{\tau_{23}}\right] d\xi. \quad (10)$$

Here $\chi_{ik} = \chi(\tau_{ik})$ are the components of the dielectric susceptibility tensor that are responsible for the formation of longitudinal ($\chi_{13}$) and transverse ($\chi_{23}$) gratings of the refractive index and appear as a result of interference of co-propagating and counter-propagating waves, respectively; $\tau_{ik}$ are the characteristic relaxation times of the gratings.

Substituting (10) into (9) and supposing for simplicity that $\tau_{13} = \tau_{23} = \tau$, we obtain the expression for the conjugated wave

$$E_4(x, t) = \frac{A}{\tau}\left\{\chi_{13}\int_0^l dx' E_2(x', t') \int_{-\infty}^{t'} E_1(x', \xi) E_3^*(x', \xi) \exp\left[-\frac{t'-\xi}{\tau}\right] d\xi + \right. \\ \left. + \chi_{23}\int_0^l dx' E_1(x', t') \int_{-\infty}^{t'} E_2(x', \xi) E_3^*(x', \xi) \exp\left[-\frac{t'-\xi}{\tau}\right] d\xi \right\}. \quad (11)$$

In order to separate out the explicit contributions of individual gratings to backward reflection of the signal, the indices of corresponding components of the dielectric susceptibility tensor are kept in (11). The following designation is used for shortening the expression

$$t' = t - |x - x'|/c. \quad (12)$$

It should be noted that for the media with fast relaxation of nonlinearity $(\tau\Delta\nu \ll 1)$, one obtains from (11)

$$E_4(x,\ t) = A\int_0^l E_1(x',\ t')\,E_2(x',\ t')\,E_3^*(x',\ t')\,dx'. \qquad (13)$$

If all the functions under the integral are reduced with the help of (8) and (12) to the temporal variation of the corresponding waves in the cross section $x = 0$, then the following expression can be obtained

$$E_4(0,\ t) = AE_2(t)\int_0^{2T} E_1(t-\tau)\,E_3^*(t-\tau)\,d\tau, \qquad (14)$$

where $T = l/c$ is the time of signal propagation through the interaction region. Relation (14) coincides with the expression for the conjugated wave in the media with instantaneous response [20] (see also [40]).

Since the reflected signal intensity is equal to $I_4 = E_4 E_4^*$, then according to (11), the average value $\langle I_4\rangle$ for the fluctuating fields can be represented in the form

$$\langle I_4\rangle = |A|^2\{|\chi_{13}|^2 S_1 + (\chi_{13}\chi_{23}^* S_2 + \text{c. c.}) + |\chi_{23}|^2 S_3\}, \qquad (15)$$

where

$$S_1 = \iint_0^l dx'\,dx''\left\langle E_2(x',\ t')\,E_2^*(x'',\ t'')\frac{1}{\tau^2}\int_{-\infty}^{t'} d\xi \times\right.$$
$$\left.\times\int_{-\infty}^{t''} d\eta\,\exp\left[(\xi+\eta-(t'+t''))/\tau\right] E_1(x',\xi)\,E_1^*(x'',\eta)\,E_3^*(x',\xi)\,E_3(x'',\eta)\right\rangle,$$

$$S_2 = \iint_0^l dx'\,dx''\left\langle E_2(x',\ t')\,E_1^*(x'',\ t'')\frac{1}{\tau^2}\int_{-\infty}^{t'} d\xi \times\right.$$
$$\left.\times\int_{-\infty}^{t''} d\eta\,\exp\left[(\xi+\eta-(t'+t''))/\tau\right] E_1(x',\xi)\,E_2^*(x'',\eta)\,E_3^*(x',\xi)\,E_3(x'',\eta)\right\rangle,$$

$$S_3 = \int_0^l\int dx'\,dx'' \Big\langle E_1(x',t')E_1^*(x'',t'')\frac{1}{\tau^2}\int_{-\infty}^{t'} d\xi \times$$
$$\times \int_{-\infty}^{t''} d\eta \exp\left[(\xi+\eta-(t'+t''))/\tau\right] E_2(x',\xi)E_2^*(x'',\eta)E_3^*(x',\xi)E_3(x'',\eta)\Big\rangle.$$

In accordance with (11), designations $t' = t - |x-x'|/c$ and $t'' = t - |x-x''|/c$ are used here again, while the corner brackets denote statistical averaging of the fields.

Restricting ourselves to the gaussian statistics of random processes, let us consider the case of uncorrelated fields $E_1$, $E_2$, and $E_3$. For example, they can be obtained from different sources (another typical case of all the three fields obtained from a single source was analyzed in [20]). If the fields are uncorrelated, then the cross-correlation function of these fields may be taken in the following form [41]:

$$\langle E_i(t')E_j^*(t'')\rangle = \langle I_i\rangle\,\delta_{ij}\exp\left[-{}^1\!/_2\Omega^2(t'-t'')^2\right], \quad (16)$$

where $\delta_{ij}$ is the Kroneker symbol and $\Omega = \Delta\omega$ is the characteristic spectral width of the interacting waves.

Since the uncorrelation of guassian random quantities results in their statistical independence, then averaging under the integral sign in (15) can be divided into the product of correlators. This circumstance considerably simplifies the calculations of $\langle I_4 \rangle$. Let us consider two the most typical cases, nonlinearities with fast and slow relaxation.

## 4.2. Inertialess nonlinearity

In the case of fast relaxation ($\tau\Delta\omega \ll 1$), one can easily find in view of (16) that $S_2 = S_2^* = 0$. Then (15) takes the form

$$\langle I_4\rangle = |A|^2(|\chi_{13}|^2+|\chi_{23}|^2)\iint_0^l dx'\,dx''\,\langle E_2(x',t')E_2^*(x'',t'')\rangle \times$$
$$\times \langle E_1(x',t')E_1^*(x'',t'')\rangle\langle E_3(x',t')E_3^*(x'',t'')\rangle. \quad (17)$$

Reducing the space-time dependences of all the wave amplitudes to the temporal form in the cross section $x = 0$, according to (8) and (12) we obtain $E_{13}(x', t') = E_{13}(t - 2x'/c)$ and $E_2(x', t') = E_2(t)$. In view of these results, (17) takes the form

$$\langle I_4\rangle = |A|^2(|\chi_{13}|^2+|\chi_{23}|^2)\langle I_1\rangle\langle I_2\rangle\langle I_3\rangle\iint_0^l e^{-\frac{4}{l_K^2}(\xi-\eta)^2}\,d\xi\,d\eta, \quad (18)$$

where $l_c = c/\Omega$ is the characteristic coherence length of radiation, i.e. the space interval in which the waves are phase-correlated. Substitititing the main variables $\xi$, $\eta$ by the new ones $x = \xi - \eta$ and $y = \xi + \eta$ (the corresponding integration domains are shown in Fig.3) and taking into account the value of Yacobian $J(\xi, \eta; x, y) = 1/2$ for the present case, one can find with a little manipulation

$$\langle I_4\rangle = |A|^2(|\chi_{13}|^2+$$
$$+|\chi_{23}|^2)\langle I_1\rangle\langle I_2\rangle\langle I_3\rangle\, l_K\left\{\frac{\sqrt{\pi}}{2}\frac{l}{l_c}\Phi(2l/l_c)+\frac{1}{4}\left(e^{-4l^2/l_c^2}-1\right)\right\}, \quad (19)$$

where $\Phi(x) = \frac{2}{\sqrt{\pi}}\int_0^x e^{-\xi^2}d\xi$ is the probability integral.

In particular cases of short ($l/l_c \ll 1$) and long ($l/l_c \gg 1$) regions of nonlinear interaction, (19) gives:

$$\langle I_4\rangle = |A|^2(|\chi_{13}|^2+|\chi_{23}|^2)\langle I_1\rangle\langle I_2\rangle\langle I_3\rangle\begin{cases} l^2, & l \ll l_c \\ \frac{\sqrt{\pi}}{2}\, l l_c & l \gg l_c \end{cases} \quad (20)$$

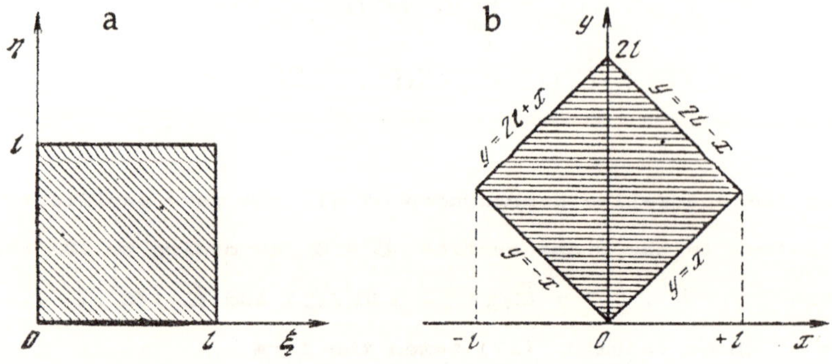


Fig.3. Integration domains in the main variables $\xi$, $\eta$ (a) and in the new variables $x = \xi - \eta$, $y = \xi + \eta$ (b) (for calculating integral (18)).

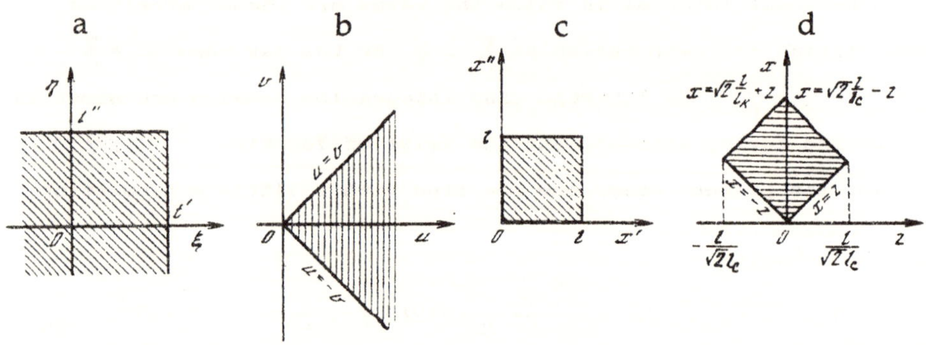


Fig.4. Integration domains in the main variables (a, b) and in the new (c, d) variables given by formula (23) (for calculating integrals (21) and (22)).

Expression (20) shows that FWI of nonmonochromatic radiation in the inertialess nonlinear media has the same efficiency as in the monochromatic case if only the length of the region of nonlinear interaction is small compared to the coherence length ( $l \ll l_c$). In this case all the parts of the interaction region are in phase.

For the long media, interpretation of the results is qualitatively different. The signal and the reference (pump) beams remain coherent only within the time $t_c = 1/\Omega$. Therefore the region of nonlinear interaction (having the length $l$ ) can be devided into $N = l/l_c$ parts that are not mutually phased. In this connection the total reflected signal is formed from the reflected signals obtained from each of the parts having the length $l_c$. It is natural that the corresponding fields add up incoherently. Summation of the intensities gives $\langle I_4 \rangle = N I_{4c} \sim (l/l_c) l_c^2 = l l_c$ (see (20)). This results in a reflection efficiency decrease.

### 4.3. Nonlinearity with slow relaxation

Let us consider the case $\Omega\tau \gg 1$, i.e. $\tau \gg 1/\Delta\omega \sim t_c$. Again, as in the section 4.2, $t_c$ is the characteristic time scale of the radiation coherence.

Taking the time interval $T$ ($t_c \ll T \ll \tau$) for averaging the fluctuations, it is possible to pull the averaging sign <...> in the expressions for $S_1$, $S_2$, and $S_3$ directly inside the time integrals. Then, in view of (16), we again have $S_2 = 0$ and obtain the following expressions for $S_1$ and $S_3$

$$S_1 = \frac{1}{\tau^2} \iint_0^l dx'\, dx'' \langle E_2(x', t') E_2^*(x'', t'') \rangle \int_{-\infty}^{t'} d\xi \times$$

$$\times \int_{-\infty}^{t''} d\eta \exp\left[(\xi + \eta - (t' + t''))/\tau\right] \langle E_1(x', \xi) E_1^*(x'', \eta) \rangle \langle E_3^*(x', \xi) E_3(x'', \eta) \rangle,$$

$$S_3 = \frac{1}{\tau^2} \int\int_0^l dx'\, dx''\, \langle E_1(x', t')\, E_1^*(x'', t'')\rangle \int_{-\infty}^{t'} d\xi \times$$

$$\times \int_{-\infty}^{t''} d\eta \exp[(\xi+\eta-(t'+t''))/\tau]\, \langle E_2(x', \xi)\, E_2^*(x'', \eta)\rangle\, \langle E_3^*(x', \xi)\, E_3(x'', \eta)\rangle.$$

Since $E_{13}(x', \xi) = E_{1,3}(\xi - x'/c)$ and $E_2(x', \xi) = E_2(\xi + x'/c)$, then, in view of (16), we obtain

$$S_1 = \frac{\langle I_1\rangle\langle I_2\rangle\langle I_3\rangle}{\tau^2} \int\int_0^l dx' dx'' \int_{-\infty}^{t'} d\xi \times$$

$$\times \int_{-\infty}^{t''} d\eta \exp[(\xi+\eta-(t'+t''))/\tau] \exp\left\{-\Omega^2\left[(\xi-\eta)+\frac{1}{c}(x''-x')\right]^2\right\} \tag{21}$$

and

$$S_3 = \frac{\langle I_1\rangle\langle I_2\rangle\langle I_3\rangle}{\tau^2} \int\int_0^l dx'\, dx'' \exp\left[-2\frac{\Omega^2}{c^2}(x'+x'')^2\right] \times$$

$$\times \int_{-\infty}^{t'} d\xi \int_{-\infty}^{t''} d\eta \exp[(\xi+\eta-(t'+t''))/\tau] \times$$

$$\times \exp\left\{-{}^1\!/_2\Omega^2\left[(\xi-\eta)+\frac{1}{c}(x'-x'')\right]^2\right\} \times \tag{22}$$

$$\times \exp\left\{-{}^1\!/_2\Omega^2\left[(\xi-\eta)+\frac{1}{c}(x''-x')\right]^2\right\}.$$

Taking the new time $(u, v)$ and space $(x, z)$ variables with the help of relations

$$\xi - t + \frac{x'}{c} = -\frac{u-v}{\sqrt{2}\,\Omega}, \qquad \eta - t + \frac{x''}{c} = -\frac{u+v}{\sqrt{2}\,\Omega}, \tag{23}$$

and

$$x' = \frac{c}{\sqrt{2}\,\Omega}(x-z), \qquad x'' = \frac{c}{\sqrt{2}\,\Omega}(x+z)$$

(the corresponding integration domains are shown in Fig.4), on rather cumbersome rearrangements, we obtain on the basis of (21)-(23)

$$S_1 = \langle I_1\rangle \langle I_2\rangle \langle I_3\rangle \sqrt{\frac{\pi}{2}} \frac{l_c^2}{\Omega\tau} \cdot \int\limits_0^{l/\sqrt{2}\, l_c} dz e^{-8z^2} \left(\frac{l}{l_c} - \sqrt{2}\, z\right) \times$$

$$\times \left\{ e^{\frac{1}{4}(1/\Omega\tau + 4\sqrt{2}\, z)^2} \left[1 - \Phi\left(\frac{1/\Omega\tau + 4\sqrt{2}\, z}{2}\right)\right] + \right.$$

$$\left. + e^{\frac{1}{4}(1/\Omega\tau - 4\sqrt{2} z)^2} \left[1 - \Phi\left(\frac{1/\Omega\tau - 4\sqrt{2}\, z}{2}\right)\right]\right\}, \qquad (24)$$

$$S_3 = \langle I_1\rangle \langle I_2\rangle \langle I_3\rangle \sqrt{\frac{\pi}{2}} \frac{l_c^2}{\Omega\tau} \int\limits_0^{l/\sqrt{2}\, l_c} dz e^{-8z^2} \left(\frac{l}{l_c} - \sqrt{2}\, z\right) \times$$

$$\times \left\{ e^{\frac{1}{4}(1/\Omega\tau + 2\sqrt{2}\, z)^2} \left[1 - \Phi\left(\frac{1/\Omega\tau + 2\sqrt{2}\, z}{2}\right)\right] + \right.$$

$$\left. + e^{\frac{1}{4}(1/\Omega\tau - 2\sqrt{2}\, z)^2} \left[1 - \Phi\left(\frac{1/\Omega\tau - 2\sqrt{2}\, z}{2}\right)\right]\right\}.$$

Since $\Omega\tau \gg 1$, the expressions obtained can be substantially simplified. Indeed, supposing that $1/\Omega\tau \ll l/l_c$ and using the relation $\Phi(-x) = -\Phi(x)$, we obtain

$$S_1 = \langle I_1\rangle \langle I_2\rangle \langle I_3\rangle \sqrt{2\pi} \frac{l_c^2}{\Omega\tau} \int\limits_0^{l/\sqrt{2}\, l_c} (l/l_c - \sqrt{2}\, z)\, e^{-\frac{\sqrt{8}\, z}{\Omega\tau}} dz =$$

$$= \frac{\sqrt{\pi}\, l_c^2}{\Omega\tau} \langle I_1\rangle \langle I_2\rangle \langle I_3\rangle \left[\frac{l}{l_c} \frac{\Omega\tau}{2} - \left(\frac{\Omega\tau}{2}\right)^2 \left(1 - e^{-\frac{2l}{l_c} \frac{1}{\Omega\tau}}\right)\right], \qquad (25)$$

$$S_3 = \langle I_1\rangle \langle I_2\rangle \langle I_3\rangle \sqrt{2\pi} \frac{l_c^2}{\Omega\tau} \int\limits_0^{l/\sqrt{2}\, l_c} (l/l_c - \sqrt{2}\, z)\, e^{-6z^2 - \sqrt{2}\, z/\Omega\tau} dz =$$

$$= \langle I_1\rangle \langle I_2\rangle \langle I_3\rangle \frac{\sqrt{2\pi}\, l_c^2}{\Omega\tau} \left\{\frac{\sqrt{\pi}}{2\sqrt{6}} \frac{l}{l_c} \Phi\left(\sqrt{3} \frac{l}{l_c}\right) + \frac{1}{6\sqrt{2}} (e^{-3l^2/l_c^2} - 1)\right\}.$$

Thus, in view of (15), (24), and (25), we finally have

$$
\begin{aligned}
\langle I_4\rangle &= |A|^2 \langle I_1\rangle \langle I_2\rangle \langle I_3\rangle (|\chi_{13}|^2 S_1 + |\chi_{23}|^2 S_3) = \\
&= |A|^2 \langle I_1\rangle \langle I_2\rangle \langle I_3\rangle \Big\{ \frac{\sqrt{\pi}\, l_c^2}{\Omega\tau} |\chi_{13}|^2 \Big[ \frac{l}{l_c}\frac{\Omega\tau}{2} - \Big(\frac{\Omega\tau}{2}\Big)^2 (1 - e^{-2e/l_c \Omega\tau}) \Big] + \\
&+ \frac{\sqrt{2\pi}}{\Omega\tau} l_c^2 |\chi_{23}|^2 \Big[ \frac{\sqrt{\pi}}{2\sqrt{6}} \frac{l}{l_c} \Phi\Big(\sqrt{3}\frac{l}{l_c}\Big) + \frac{1}{6\sqrt{2}} \big(e^{-3l^2/l_c^2} - 1\big) \Big] \Big\}.
\end{aligned} \tag{26}
$$

We note that in the case of short nonlinear media ($l/l_c \ll 1$), (24) and (25) give $S_3 \to S_1 = \sqrt{\pi} l^2/2\Omega\tau$. Therefore

$$\langle I_4\rangle = |A|^2 \langle I_1\rangle \langle I_2\rangle \langle I_3\rangle (|\chi_{13}|^2 + |\chi_{23}|^2) \sqrt{\pi}\, l^2/2\Omega\tau.$$

Since $\Omega\tau \gg 1$, then comparing this expression with (20), we see that the reflection efficiency of nonmonochromatic radiation in the media with slow relaxation substantially decreases (with respect to the inertialess case).

Further, in the limiting case of a long medium ($l/l_c \gg 1$), the expression for $S_1$ has two limiting values for opposite relations between the relaxation time $\tau$ and the transit time of a pulse in a nonlinear medium. When $\Omega\tau \gg l/l_c \gg 1$, we have $S_1 = (\sqrt{\pi} l^2/2\Omega\tau) \langle I_1\rangle \langle I_2\rangle \langle I_3\rangle$ as in the case of a short medium. For $l/l_c \gg \Omega\tau \gg 1$, expression (24) takes the form

$$S_1 = \langle I_1\rangle \langle I_2\rangle \langle I_3\rangle \sqrt{\pi}\, l l_c/2.$$

In both cases, (5) gives the following expression for $S_3$:

$$S_3 = \pi/\sqrt{12}\, \frac{l l_c}{\Omega\tau} \langle I_1\rangle \langle I_2\rangle \langle I_3\rangle \ll S_1.$$

This means that the transverse grating formed by variation of the refractive index does not give any appreciable contribution to the reflection of nonmonochromatic radiation.

Finally we can write

$$\langle I_4\rangle = |A|^2 \langle I_1\rangle \langle I_2\rangle \langle I_3\rangle \frac{\sqrt{\pi}}{2} \times$$

$$\frac{l^2}{\Omega\tau}\left(|\chi_{13}|^2 + |\chi_{23}|^2\right) \quad \text{for} \quad l/l_c \ll 1,$$

$$\frac{l^2}{\Omega\tau}\left(|\chi_{13}|^2 + \sqrt{\frac{\pi}{3}}\,|\chi_{23}|^2 \frac{l}{l_c}\right) \quad \text{for} \quad \Omega\tau \gg \frac{l}{l_c} \gg 1, \qquad (27)$$

$$ll_c\left(|\chi_{13}|^2 + \sqrt{\frac{\pi}{3}}\,|\chi_{23}|^2 \frac{1}{\Omega\tau}\right) \quad \text{for} \quad \frac{l}{l_c} \gg \Omega\tau \gg 1.$$

Interpretation of the results obtained is similar to interpretation of relations (20). In the case under consideration, the pulse trains of the signal and the co-propagating reference wave with the duration $t_c = 1/\Delta\omega$ record in a nonlinear medium a grating formed by refractive index variations. The experiments on self-diffraction [36, 42] showed that the efficiency of this record in a medium with relaxation decreases approximately by a factor of $(t_c/\tau = 1/\Omega\tau)$ in comparison with the stationary case. The recorded grating is present in the medium within the time $\tau$. The length of a coherently recorded part of the grating is estimated as $c\tau$ in our case. Therefore, this grating has the reflection efficiency of the order of $(c\tau)^2 t_c \tau$. If the interaction region is sufficiently long (the last condition in (27)), then the number of incoherently summed gratings is equal to $N = l/c\tau$. Therefore, the FWI efficiency is of the order of $((c\tau)^2 t_c/\tau)N = ll_c$ in this case.

Comparing the interaction efficiencies of incoherent waves in the media with fast and slow relaxation of nonlinearity (for $l \ll l_c$), we find that in the first case the FWI efficiency is higher by a factor of $\Omega\tau$ than in the latter case.

## 5. EXPERIMENTAL INVESTIGATION OF THE EFFICIENCY AND THE QUALITY OF PHASE CONJUGATION IN THE PROCESS OF FOUR-WAVE INTERACTION OF NONMONOCHROMATIC RADIATION

In this section [38, 43], most attention is given to experimental investigation of the effect of phase matching of interacting waves on the reflection efficiency. The dependence of the angular spectrum of the reflected wave on the interaction geometry is also studied.

### 5.1. Description of the experimental setup

A wideband laser oscillating on neodymium glass GLS-1 was used in the experiments. When choosing the laser, the task of obtaining the widest possible radiation spectrum with a sufficient intensity was set. The oscillators with active Q-sitching satisfy this requirement. Lasing with a spectral width of 20 $cm^{-1}$ at half maximum is easily obtainable in this system. In the similar oscillators with passive Q-switching, the laser spectrum is usually no more than several $cm^{-1}$. This is explained by the fact that the number of passes of radiation through the laser cavity preceeding the oscillation threshold is considerably greater in the latter case. It should be also mentioned that the laser with active Q-switching is in commercial production and is often used in practice.

A schematic diagram of the multifrequency Nd laser with active Q-switching is given in Fig.5. Q-switching was performed by a Kerr cell C positioned between the Glan prisms $P_1$ and $P_2$ oriented at an angle $45^{\circ}$ to the electrodes of the cell. $TEM_{00}$ oscillation mode was provided by the size of a diaphragm D. Fig.5 shows that almost all the optical elements (besides the active element having the facets cut at a $2^{\circ}$ angle) introduce selective losses into the laser cavity.

A typical spectrogram of an oscillator pulse is presented in Fig.6a. It was recorded by a Fabry-Perot interferometer with a 0.08 cm spacing that gave a 62.5 $cm^{-1}$ free spectral range. Fig.6b shows the microdensitogram of a spectrum recorded on a I 1060V film and photometered using a 500 x 1000 μm slit A. The temporal dependence of laser radiation and the corresponding spectral distribution varied from shot to shot. This is typical for radiation of multifrequency lasers with a noncorrelated distribution of longitudinal modes [44]. However, the typical scale of spectrum modulation (6 $cm^{-1}$) and its half-width (15-20 $cm^{-1}$) had small variations.

Radiation of the multifrequency oscillator having a diffraction divergence $5 \cdot 10^{-4}$ rad was amplified in a two-pass amplifier and then after a Faraday shutter entered a setup for FWI investigation (Fig.7). A No.3955 dye solution in nitrobenzene having the relaxation time of the order of 40 ps was used as a nonlinear medium [45]. It was poured into a cell C 0.165 cm thick. The solution concentration was chosen to have approximately 50% small-signal transmittion of the cell at the operating wavelength. Nearly 20% of radiation was separated by a dielectric mirror $M_1$, passed through a diaphragm D 1 mm in diameter and then was directed by a 50% mirror $M_3$ into the cavity C forming a signal wave in it. The reference waves were produced by splitting radiation into two equal parts by the mirror $M_2$. Using rotation of the mirrors $M_4$ amd $M_5$, the coincidence of reference beams in the cell to an accuracy of $5 \cdot 10^{-4}$ rad was obtained. The reference beams were approximately 3 mm in diameter. Their intensity was varied in the range 2-20 $MW/cm^2$ depending on the amplifier pump level. The signal wave intensity was equal to 20% of the reference wave intensities.

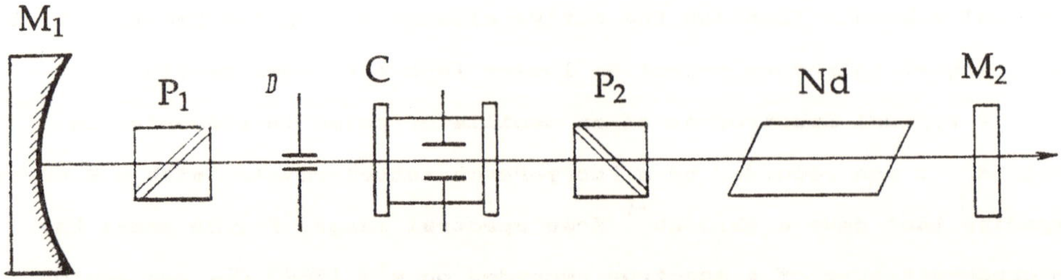


Fig. 5. Schematic diagram of the multifrequency Nd-laser with active $Q$-switching:
Nd - active element of GLS-1 glass 260 mm in length; $M_1$ and $M_2$ - dielectic mirrors; $P_1$ and $P_2$ - crossed Glan-Thomson polarizers; C - Kerr cell; D - 2 mm diameter diaphragm.

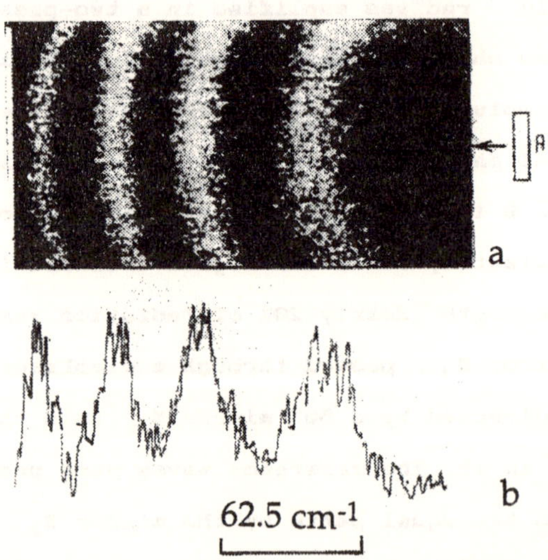


Fig. 6. Photograph of the laser radiation spectrum recoreded by a Fabri-Perot interferometer (a) and the corresponding densitogram (b). The densitogram accuracy (spectral scale is given) is governed by slit A.

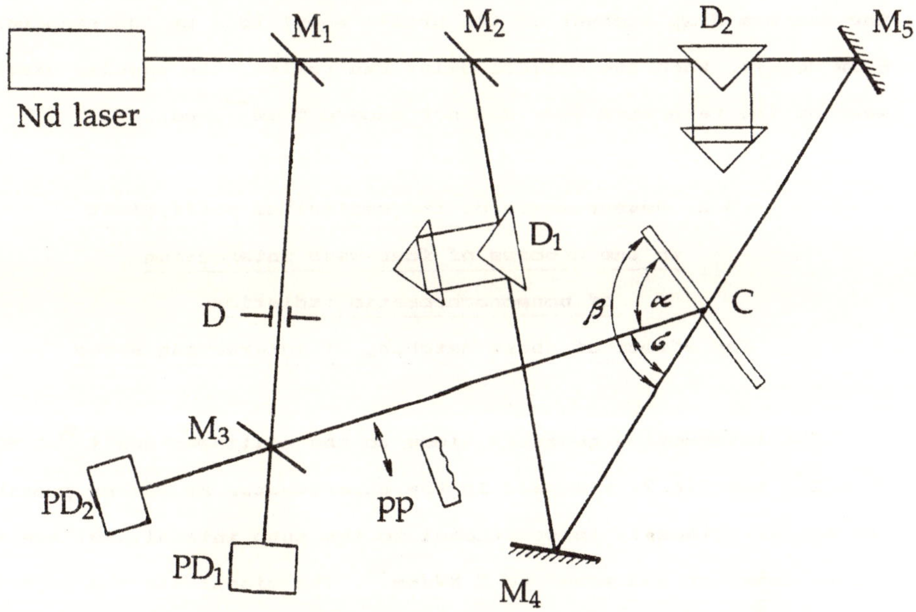


Fig. 7. Experimental setup for investigating the characteristics of FWI interaction of nonmonochromatic waves:

$M_1$ – $M_5$ – dielectric mirrors, $PD_1$ and $PD_2$ – detectors for signal and reflected waves, $D_1$ and $D_2$ – optical delays, C – cell, D – diaphragm, PP – phase plate, $\sigma$ – angle of convergence of the signal and the reference waves, angle $\alpha$ is additional to the incidence angle of the signal wave, $\beta \leq \alpha + \sigma$.

The optical delays $D_1$ and $D_2$ were used for synchronizing all the three waves in the interaction region. They provided the synchronization accuracy better than 0.02 mm. The delay lines were aligned in a way to have the relative misalignment of reference beams less than 0.5 mm at the maximum displacement of the prisms equal to 3 mm (corresponding to 6 cm delay). When the optical delay was varied, the angular displacement of the reference wave did not exceed $5 \cdot 10^{-4}$ rad.

### 5.2. Measurements of the reflection coefficient in the process of four-wave interaction of nonmonochromatic radiation. The effect of phase matching of interacting waves.

The interaction geometry close to the collinear one ($\beta \simeq 90^\circ$ and $\delta \simeq 2^\circ$, see Fig.7) was used in the experiments. As far as possible, the signal intensity (proportional to the pump intensities) was maintained constant and equal to 2 $MW/cm^2$. The signal and the reference radiation pulses were integrated by FD 9E111 photodiodes $PD_1$ and $PD_2$. The corresponding amplitudes were recorded by a two-channel storing S8-14 oscilloscope. Since the pulse shape and the intensity of the signals as well as of the reference waves were not reproduced from shot to shot, it was necessary to measure the dependence of the signal reflection coefficient (for energy) on the energy of the reference waves. As expected, this dependence appeared to be quadratic for the given intensity to an accuracy of the 20% statistical mean spread. We note that such a spread was absent in the experiments with a single--frequency oscillator used for investigation of FWI in the same dye solution [38]. Hence the reflection coefficient was normalized by the quadratic law with variation of the energy level. The relative error of this normalization was determined by the statistical spread.

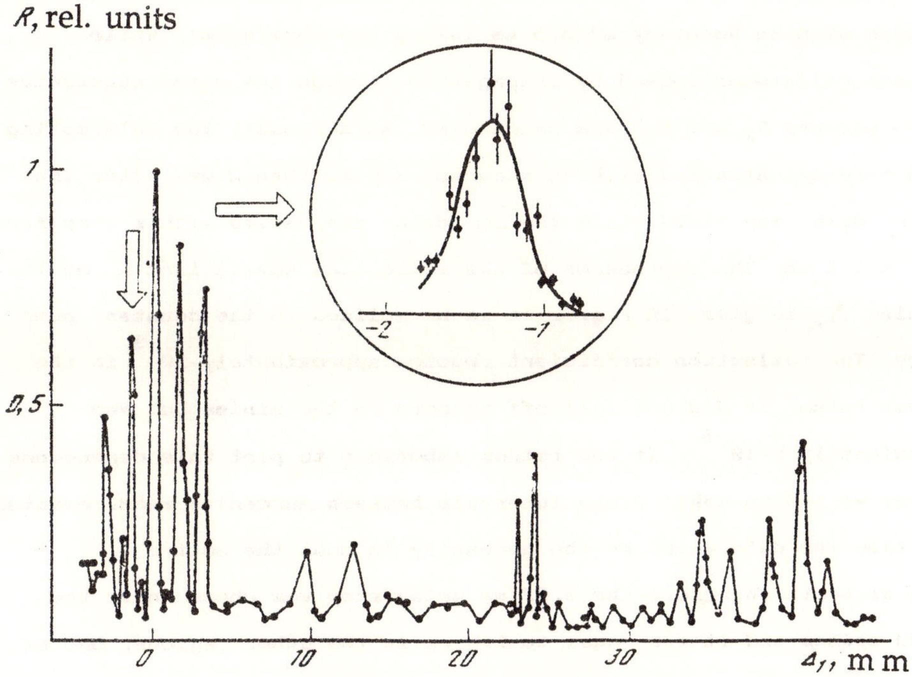


Fig. 8. Dependence of the reflection coefficient on the detuning of co-propagating interacting waves.

The optical paths of all the three interacting waves were initially synchronized with the best possible accuracy. For this purpose all the optical paths in air starting from the mirror $M_1$ were measured with an accuracy of 0.5 mm (along the beam axis), while the path difference caused by propagation through the glass substrates of the mirrors $M_1$ and $M_2$ was calculated. As a result, the interacting beams were synchronized with an accuracy better than 2 mm. After that the $D_2$ delay was fixed while the $D_1$ delay was varied with a step from 0.06 to 1.0 mm. The dependence of the reflection coefficient $R$ on the detuning $\Delta_1$ is given in Fig.8. It is normalized to the constant pump energy. The reflection coefficient reached approximately $10^{-3}$ in the maximum spike. It did not fall off to zero in the minima but was approximately $5\cdot10^{-5}$. It was rather laborious to plot this dependence because we had to make 2 min intervals between succesive measurements. This time was determined by the necessity to cool the active laser element. Therefore the 0.02 mm delay step was chosen near the marked spikes and it was equal to 1.0 mm in the other regions. One of the highest peaks of the reflection coefficient was recorded in all the possible detail using 0.06 mm delay step. To expose the typical structure of reflection peaks, this dependence is given separately in Fig.8 in a different scale. It is necessary to note that the $D_2$ delay variation had no considerable effect on the reflection coefficient neither in the reflection maxima nor in the regions, where the reflection coefficient was practically independent of $\Delta_1$ ($\Delta_1 >$ 40 mm).

In the case under consideration, the relaxation time of the dye solution nonlinearity ($\tau \simeq 4\cdot10^{-11}$ s [45]) was considerably greater than the time of characteristic fluctuations of radiation, which is estimated as $1/\Delta\omega \simeq 2.6\cdot10^{-13}$ s (the spectral width is $\Delta\nu$ = 20 cm$^{-1}$). Therefore the nonlinear medium behaved as the inertial one ($\Delta\omega\tau \simeq$ 160). The coherence length of radiation $l_c = c/\Delta\omega \simeq 8\cdot10^{-3}$ cm was considerably less than the interaction length ( $l/l_c \simeq 20$). At this $l/l_c$ ratio, only the longitudinal grating of the optical density variation could give contribution to reflection. In this case the reflected wave intensity in the region of partial coherence of the waves $E_1$ and $E_2$ is proportional to the square of the modulus of the correlation function of these fields. It can be shown that in the case of collinear geometry, the correlator $< E_1 E_3^* > = f(t_0)$ is the Fourier-transform of the radiation spectrum.

Hence the function shown in Fig.8 characterizes basically the Fourier-spectrum of wideband radiation of the given laser. The radiation spectrum is unambiguously related to the Fourier-spectrum. It is interesting to note that the spectrum shape varied from shot to shot, while the Fourier-spectrum detected in a hundred of shots slowly varied with small $\Delta_1$ displacements. For $\Delta_1$ > 15 mm, the Fourier-spectrum reproducibility became considerably worse. This seeming discrepancy of the spectra is attributed to the stability of the substructure of the temporal spectrum which was observed in every pulse. Most likely the spectral substructure, which is unambiguously related to the Fourier-spectrum shape, is due to multiply repeated reflections of radiation from the interfaces inside the laser cavity. Let us determine the correspondence between the characteristic features of the two types of spectrum. Sharp reflection peaks following with an interval of approximately 1.55 mm are unambiguously related to the laser spectrum modulation $1/\Delta_1$ = 6.5 $cm^{-1}$ and also $1/2\Delta_1$ . The modulations of this scale are present in all the radiation spectrograms (see Fig.6). A detailed Fourier--spectrum of one of the peaks has a smooth envelope modulated with a $1.6 \cdot 10^{-2}$ cm period. The envelope is well aproximated by the function $R \sim \exp(-\Delta_1^2/\Delta_0^2)$, where $\Delta_0 \sim 2.65 \cdot 10^{-2}$ cm. The spectral distribution $\exp(-1.37 \cdot 10^{-2}\, \Delta\nu^2)$ corresponding to this envelope has a

15 $cm^{-1}$ width at half maximum. This is a good agreement with the densitograms of the laser spectrum. One can see from the estimates that frequency modulation may be caused by a spectral spike displaced by $\Delta\nu \sim 60$ $cm^{-1}$ from the main frequency and having a spectral power by an order of magnitude smaller than the latter. However, this radiation as not detected because of its low intensity.

We note that interpretation of the reflection versus delay dependence as the laser Fourier-spectrum holds only in the region of partial coherence of the reference and the signal waves, i.e. in the region of maximum reflections. At large detunings of the interacting waves their coherence is disturbed, nevertheless the reflection coefficient is nonzero. This behavior is no longer described by the Fourier-spectrum and can be interpreted on the basis of the results of section 4. According to formula (27), the ratio of the peak intensity of the reflected signal to the background is $I_{peak}/I_b \simeq \Delta\omega\tau \simeq 160$, while in the experiments this quantity was approximately $15 \pm 5$. This discrepancy is explained by the difference of simplified expression (16) used for derivation of (27) from the actual shape of Fourier--spectrum. Expression (16) corresponds to correlation of an ideal noise spectrum. The presence of $N$ similar correlation peaks at various nonzero $\Delta t$ will cause the corresponding N-fold increase of the effective length and of the coherence time. This results in an N-fold increase of the background level corresponding to FWI of incoherent fields.

### 5.3. Investigation of the angular spectrum of reflection and its effect on the phase conjugation quality

The angular spectrum of reflection was studied using the experimental setup shown in Fig.7. The photodiode $PD_2$ was replaced by a camera mounted in a plane conjugated to the plane of diaphragm ( ~1 mm in diameter) with respect to the mirror $M_3$ . A phase plate positioned at some distance from the cell was used to examine the phase properties of the reflected signal and also the possibility of compensating the phase distortions for the double pass in the configuration of PC using FWI of nonmonochromatic waves.

The angular characteristics of reflected radiation were measured at various angles of convergence $\delta = \beta - \alpha$ in the range 2-49°. Fig. 9 shows the images of the reflected beam spots recorded at $\delta = 2°$ ( $\alpha = 88°$, $\beta = 90°$ ) and $\delta = 49°$ ( $\alpha = 70°$, $\beta = 119°$ ). It is seen that in the case $\delta = 2°$ the image of the diaphragm is not practically blurred, while at $\delta = 49°$ an evident blurring of the image of the reflected beam is observed. It is symmetric with respect to the axis of the beam, which is reflected exactly backward, in the plain containing the beam axes of the interacting waves.

The intensity distributions of the reflected beam measured in the plane of interacting beams at the angles 2, 33, and 49° are shown in Fig. 10. When $\delta$ is small, the diaphragm image produced by the reflected beam is practically identical to the intensity distribution of the incident wave directly behind the diaphragm. When $\delta$ is increased, the diaphragm image extends in the plane of convergence of the signal and the reference waves. Curve 1 in Fig. 1 practically coincides with the undisturbed distribution, which corresponds to the ideal PC image.

Using the experiments with the phase plate PP positioned close to the cell C , it was possible to make sure that (curve 2) in the monochromatic case FWI-reflection had the same phase-conjugate character as in the degenerate case. When the phase plate was positioned at a distance $L \sim 2$ cm, it had no effect on the intensity distribution of reflected radiation. This indicates that FWI of nonmonochromatic waves is accompanied by PC with angular spreading. A similar phenomenon of PC accompanied by rotation of the reflected wave was observed in

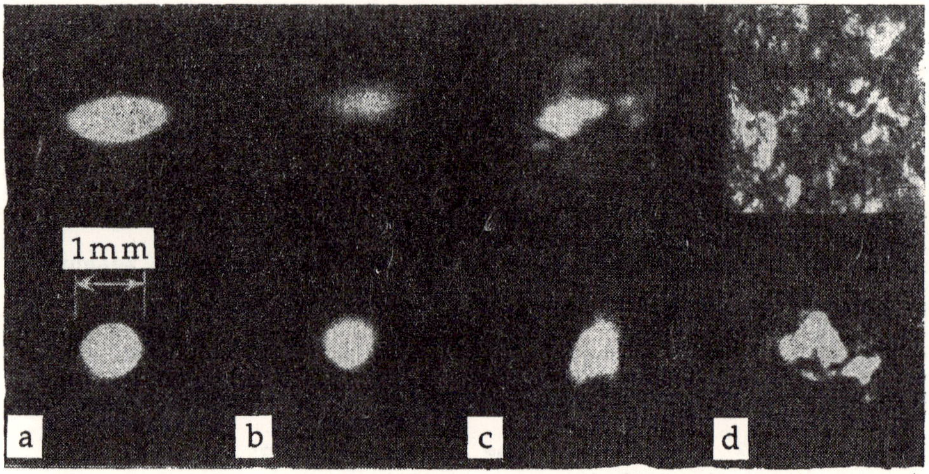


**Fig. 9. Diaphragm images produced by the reflected beam at different angles of convergence of the signal and the pump waves $\sigma = 49^{\circ}$ (above) and $2^{\circ}$ (below) without the phase plate (a) and with the phase plate at a distance of 2 (b), 20 (c), and 60 (d) cm from the cell.**

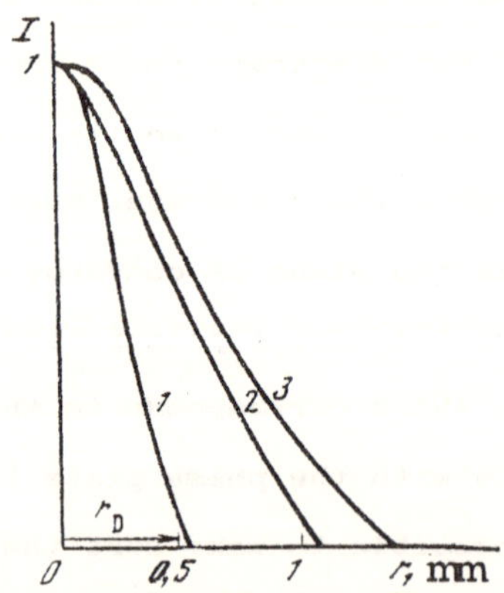


**Fig. 10. Intensity distributions of reflected radiation at different convergence angles of the interacting waves: $\sigma$ = 2 (1), 33 (2), and $49^{\circ}$ (3); $r_D$ is the diaphragm radius.**

[46, 47] when collinearity of the reference waves was disturbed.

It is obvious that both the detected angular spreading and the rotation should cause deterioration of PC in the process of FWI of nonmonochromatic radiation. The most commonly used method of the PC quality examination consists in reconstruction of the initial distribution of laser radiation with diffraction divergence after a double pass through a phase-distorting medium. The maximum angle of the reflected beam spread $\theta_{max}$ for which compensation of phase shifts obtained by the signal wave after passing the phase plate with a typical transverse dimension of inhomogenieties $\rho_x$ still takes place, can be estimated by the formula $\theta_{max} \simeq \rho_x / L$, where $L$ is the distance between the phase plate and the PC mirror. The glass phase plate with $\rho_x \sim 0.3$ mm was used in the experiments. It increased the signal beam divergence from $10^{-3}$ to $3 \cdot 10^{-3}$ rad. Its distance from the cell with a nonlinear medium varied from 2 to 60 cm. The diaphragm images produced by reflected radiation at different angles of convergence $\sigma$ and different $L$ are given in Fig. 9. As noted above, at $L = 2$ cm and the angles $\sigma$ from 2 to 49° (see Fig. 9b), full compensation of inhomogenieties to the state shown in Fig. 9a takes place. At $L = 60$ cm and $\sigma = 49°$ (see Fig. 9d above) phase inhomogenieties are not compensated at all. In the intermediate case (see Fig. 9c and Fig. 9d below) only partial compensation is observed.

On the basis of the experiments performed we can draw a conclusion that in the general case widening the frequency spectrum of the interacting waves in the FWI process results in angular spreading of the reflected wave and deterioration of the PC quality. This effect is minimized at minimum angles of convergence of the signal and one of the reference waves and in the case when the interacting waves are incident on the nonlinear medium close to its normal.

## 6. CONCLUSION

Let us formulate some conclusions and recommendations.

1. If an appropriate medium with a fast nonlinearity following the field fluctuations is taken for traditional configurations of PC with the interacting waves obtained from a single source, a four-fold increase of the reflection coefficient in comparison with a medium having a slow nonlinearity (with the same susceptibility) can be obtained. It is well known that the nonlinear susceptibility $\chi^{(3)}$ of various media has a general tendency to increase proportional to the nonlinearity relaxation time $\tau$. Since the variations of $\chi^{(3)}$ and correspondingly of $\tau$ cover the range of some orders of magnitude, it is advisable to use nonlinear media with a slow response and large susceptibility coefficient in order to obtain effective reflection of radiation in the FWI process.

2. In the case of FWI of incoherent waves the interaction efficiency is inversly proportional to the relaxation time $\tau$ of a medium. Choosing a medium with small $\tau$, with other conditions being equal, it is possible to increase the reflection efficiency up to the point in which $\tau$ becomes equal to the optical propagation time through a nonlinear medium ($l/c$). Further optimization of the medium for $\tau$ is inexpedient since the reflection coefficient will depend only on $|\chi|^2 l l_c$.

3. In the noncollinear case, when the angle of convergence of the signal and one of the reference waves is increased, this can result in the decrease of the reflection coefficient and in the spread of the reflected wave angular spectrum (corresponding to PC deterioration). It is necessary to satisfy the requirements for small value of the convergence angle $\sigma$ to obtain high PC efficiency in the process of FWI of nonmonochromatic waves. However, very small angles are inexpedient because the volume grating must be formed in the medium.

## REFERENCES

1. Zel'dovich B.Ya., Popovichev V.I., Ragul'skii V.V., Faizullov F.S. "On the relation between the wavefronts of reflected and exciting optical signals in stimulated Brillouin scattering" Pis'ma v ZhETF, 1972, vol.15, No.3, pp.160-164.

2. Denisyuk Yu.N. "On the reproduction of optical properties of an object in the wave field of scatterd radiation" Optika i spektroskopiya, 1963, vol.15, No.4, pp.522-532.

3. Kogelnik H.W. "Controlled transmission of waves through inhomogeneous media" Pat. 3449577 US, Filed 23.10.65.

4. Cathey W.T. "Holographic simulation of compensation for atmospheric wavefront distortion" Proc. IEEE, 1968, vol.56, No.3, p.340.

5. Stepanov B.E., Ivakin E.V., Rubanov A.S. "On recording of plane and volume dynamic holograms in bleachable media" DAN SSSR, vol.196, No.3, pp.567-569.

6. Woerdman J.P. "Formation of transient free carrier hologram in Si" Opt. Commun., 1970, vol.2, No.5, pp.212-214.

7. Efimkov V.F., Zubarev I.G., Kotov A.V. et al "Time response of stimulated Brillouin scattering and zero-threshold reflectivity of short pulses in a phase conjugation mode" ZhETF, 1979, vol.77, No.2, pp.526-536.

8. Bergman E.E., Bigio I.T., Feldman B.T., Fisher R.A. "High-efficiency pulsed 10.6 μ phase conjugate reflection via degenerate four-wave mixing" Opt. Lett., 1978, vol.3, No.3, pp.82-84.

9. Efimkov V.F., Zubarev I.G., Mikhailov S.I. et al. "Polarization methods of the improvement of energy characteristics of phase conjugate mirrors" Kvantovaya elektron., 1984, vol.11, No.2, pp.303-310.

10. Zel'dovich B.Ya., Pilipetskii N.F., Shkunov V.V. "Obrashchenie volnovogo fronta" [Phase conjugation] Moscow: Nauka, 1985, 252 pp.

11. Bespalov V.I., Pasmanik G.A. "Nelineinaya optika i adaptivnye lazernye systemy" [Nonlinear optics and adaptive laser systems] Moscow: Nauka, 1986, 134 pp.

12. "Nonlinear optical phase conjugation" Opt. Eng., 1982, vol.21, No.2, pp.155-284.

13. Special issue on dynamic gratings and four-wave mixing. IEEE J. Quant. Electron., 1986, vol.22, No.8, pp.1194-1543.

14. Yariv A. "Phase conjugate optics and real time holography" Ibid., 1978, vol.14, No.9, pp.650-661.

15. Gyulamiryuan A.L., Zel'dovich B.Ya., Mamaev A.V. et al. "Four-wave phase conjugation with selection and signal control" Phase conjugation of radiation in nonlinear media. Gor'kii: IPF AN SSSR, 1982, pp.91-110.

16. Yariv A., Peper D.M. "Amplified reflection, phase conjugation, and oscillation in degenerate four-wave mixing" Opt. Lett., 1977, vol.1, No.1, pp.16-18.

17. Wandzura S.M. "Effects of atomic motion on wavefront conjugation by resonantly enhanced degenerate four-wave mixing" Ibid., 1979, vol.4, No.7, pp.208-210.

18. Departie D., Haueisen D. "Multiline phase conjugation at 4μm in germanium" Ibid., 1980, vol.5, No.6, pp.252-254.

19. Steel D.G., Lam J.F. "Multiline phase conjugation in resonant materials" Ibid., No.7, pp.297-299.

20. Basov N.G., Zel'dovich B.Ya., Kovalev V.I. et al. "Reflection of multifrequency signal via four-wave interaction in germanium at 10.6 μm" Kvantovaya elekton., 1981, vol.8, No.4, pp.860-864.

21. Pepper D.M., Abrams R.L. "Narrow optical bandpass filter via nearly degenerate four-wave mixing" Opt. Lett., 1978, vol.3, No.6, pp.212-214.

22. Nilsen J., Yariv A. "Nearly degenerate four-wave mixing applied to optical filters" Appl. Opt., 1979, vol.18, No.2, pp.143-145.

23. Martin G., Hellwarth R.W. "Infrared-to-optical image conversion by Bragg reflection from thermally induced index gratings" Appl. Phys. Lett., 1979, vol.34, No.4, pp.371-374.

24. Blombergen N. "Nelineinaya optika" [Nonlinear optics] Moscow: Mir, 1966, 424 pp.

25. Garret M.H., Hoffman H.G., Karr T.J. "Quality of phase-conjugation in nondegenerate four-wave mixing" CLEO'82: Digest of technical papers. Wash. (D.C.), 198, p.166.

26. Gyulamiryan A.L., Zel'dovich B.Ya., Mamaev A.V. et al. "Tunable nonlinear four-wave filter" Optika i spektroskopiya, 1982, vol.42, No.3, pp.387-389.

27. Pilipetsky N.F., Shkunov V.V. "Narrowband four-wave reflecting filter with frequency and angular tuning" Opt.Comm., 1981, vol.37, No.3, pp.217-220.

28. Denisyuk Yu.N. "Features of phase conjugation by a three-dimensiona Doppler dynamic hologram" ZhTF, 198, vol.52, No.7, pp.1338-1345.

29. Denisyuk Yu.N. "Features of phase conjugation" in "Opticheskaya golografiya" [Optical holography] Leningrad: Nauka, 1983, pp.4-14.

30. Denisyuk Yu.N. "On the imaging properties of the running waves of intensity under recording of dynamic volume holograms" ZhTF, 1974, vol.44, No.1, pp.131-137.

31. Hellwarth R.W. "Generation of time-reversed wave fronts by nonlinea refraction" J. Opt. Soc. Amer., 1977, vol.67, pp.1-3.

32. Gyulamiryan A.L., Mamaev A.V., Pilipetskii N.F. et al. "Investigation of the nondegenerate four-wave interaction efficiency" Kvantovaya elektron., 1981, vol.8, No.1, pp.196-197.

33. Bespalov B.I., Betin A.A., Pasmanik G.A., Shilov A.A. "Phase conjugation via Raman transformation of a Stokes wave in a field of

counter-propagating pump beams" Pis'ma v ZhTF, 1979, vol.5, No.4, pp.242-246.

34. Erokhin A.I. "Phase conjugation via nondegenerate four-wave interaction and active Raman spectroscopy" Dissertation for Candidate's Degree of Physics and Mathematics, Moscow, 1988, 151 pp.

35. Erokhin A.I., Stepanov A.A., Sheglov V.A. "On the efficiency of four-wave interaction of nonmonochromatic radiation in the media with relaxation" Preprint FIAN, No.47, Moscow, 1988, 15 pp.

36. Masalov A.V. "Dynamic holography in the technique for measuring relaxation time of the media" Trudy FIAN, 1982, vol.131, pp.127-148.

37. Vladimirov V.S. "Uravneniya matematicheskoi fiziki" [Equations of mathematical physics] Moscow: Nauka, 1967, 324 pp.

38. Erokhin A.I., Morachevskii N.V., Faizullov F.S. "Reflection efficiency in four-wave incoherent interaction" Preprint FIAN, No.150, Moscow, 1980, 14 pp.

39. Smith P.W., Tomlinson W.J., Eilenberger D.J., Maloney P.J. "Measurement of electronic optical Kerr coefficients" Opt. Lett., 1981, vol.6, No.12, pp.581-583.

40. Basov N.G., Kovalev V.I., Musaev M.A., Faizullov F.S. "Phase conjugation of a pulsed $CO_2$ -laser radiation" Trudy FIAN, 1986, vol.172, pp.116-179.

41. Akhmanov S.A., D'yakov Yu.E., Chirkin A.S. "Vvedenie v statisticheskuyu radiofiziku i optiku" [Introduction to statistical radiophysics and optics] Moscow: Nauka, 1981, 640 pp.

42. Vasil'eva M.A., Malyshev V.I., Masalov A.V. "Technique for measurement of picosecond relaxation times of bleachable media using stochastic beams of laser radiation" Kratkie soobshch. po fizike, 1980, No.1, pp.35-39.

43. Erokhin A.I., Kovalev V.I., Mikheev P.A., Faiyullov F.S. "On the quality of PC of multifrequency radiation via four-wave interaction" Kvantovaya Elektron., 1985, vol.12, No.1, pp.186-188.

44. Malyshev V.I., Masalov A.V., Milanich A.I. "Energy distribution of a neodimium laser over the modes" Ibid., 1975, vol.2, No.9, pp.1963-1968.

45. Babenko A.A., Dyadyusha G.G., Kudinova M.A. et al. "New compounds for passive shutters of the near-IR lasers" Ibid., 1980, vol.7, No.8, pp.1796-1802.

46. Bloom D.M., Bjiorklund G.C. "Conjugate wave-front generation and image reconstruction by four-wave mixing" Appl. Phys. Lett., 1977, vol.31, No.9, p.592-594.

47. Blashchuk V.N., Zel'dovich B.Ya., Mamaev A.V. et al. Phase conjugation with a rotation - theory and experiment for four-wave interaction" Nonlinear optics: Proc. VI Vavilov conf., Novosibirsk, 1979. Novosibirsk: Institute for thermal physics , SO AN SSSR, 1979, pt.2, pp.197-202.

# SUBJECT INDEX